Pre-Calculus and Trig Help

By Kathryn Paulk
Copyright © 2024

Updated: 09/01/2024

Table of Contents

Introduction 1

Algebra Topics – Part 1 3

   Polynomial Division (Long & Synthetic) 4

   Factoring Polynomials 10

   Why Factor Polynomials 21

   Completing The Square 23

   Quadratic Formula 28

   Function Composition 38

   Inverse Functions 43

   Transformations 49

   Graphing Quadratic Functions 65

   Graphing Polynomial Functions 77

   Graphing Rational Functions 84

   Working With Radicals 100

   Logarithms 113

   Exponential Growth and Decay 128

   Regression 135

Systems of Linear Equations 152
Systems of Linear Inequalities 158
Solving Inequalities 162

Algebra Topics – Part 2 .... *170*

Roots 171
Binomial Expansion 187
Partial Fractions 195
Conics 206
Vectors 215
Parametric Equations 221
Series 230
Matrices – Basic Operations 240
Matrices – Systems of Eqns. 247
Matrices – More Examples 257
Statistics 271
Probability 284
Limits 301
Difference Quotient (Derivative) 330

Trig Topics – Part 1 342
Right Triangles 343
The Unit Circle...................................... 350
Trigonometric Functions 359
Trig Functions of General Angles 368
Law of Cosines 377
Law of Sines 381
General Triangles 386
Bearings.. 398

Trig Topics – Part 2 402
Graphs of Trig Functions........................... 403
Graphs of Inverse Trig Functions 407
Translations of Trig Functions..................... 410
Trig Identities....................................... 418
Trig Formulas 424
Using Trig Identities 428
Solving Trig Equations 440
Using Trig Formulas................................. 450

Trig Topics – Part 3 .... *463*
   Complex Numbers – Rectangular Form 464
   Complex Numbers – Polar Form 467
   Complex Numbers – Conjugate Pairs 474
   Complex Numbers – Operations 478
   Complex Numbers – DeMoivre's Thrm. 484
   Complex Numbers – Roots 490
   Polar Curves 499
   Polar Curves – Patterns 513
   Polar Curves – More Examples 524

References *545*
Other Books by Kathryn Paulk *547*

Introduction

This book will help students who are currently taking or planning to take a course in or <u>Pre-Calculus and Trig</u>. For each topic, key equations are listed and followed by detailed examples.

This book has been formatted so that it is easy to read on both paperback and also on electronic devices with the Kindle app (laptop, iPad, Kindle E-reader, and iPhone).

Please note that four "Help" books with different titles contain overlapping content.

Book Title	Book Content					
	Alg. 0	Alg. 1	Alg. 2	Trig. 1	Trig. 2	Trig. 3
Algebra 1 Help	X	X		X		
Algebra 2 Help		X	X	X	X	X
Pre-Calc. & Trig. Help		X	X	X	X	X
College Algebra Help	X	X	X	X	X	

Algebra Topics – Part 1

Polynomial Division (Long & Synthetic)

Division -- Review

$$\frac{Dividend}{Divisor} = Quotient \rightarrow \text{Divisor} \overline{\smash{)}\text{Dividend}}^{\text{Quotient}}$$

Regular Division with Numbers (Quick Review)

$\dfrac{425}{5}$

$= 85$

```
        8 5
    5 ) 4 2 5
        4 0
        ─────
          2 5
          2 5
          ─────
            0   Remainder
```

$\dfrac{426}{5}$

$= 85 + \dfrac{1}{5}$

```
        8 5
    5 ) 4 2 6
        4 0
        ─────
          2 6
          2 5
          ─────
            1   Remainder
```

Division of Polynomials

- **Long division** of polynomials is similar to regular division. No restrictions on the divisor.
- **Synthetic division** is only used with 1st degree divisors in the form $(x - c)$.

For comparison, the same problem is done with long and synthetic division, below.

$$\frac{x^3 + 3x^2 - x - 3}{x - 1}$$

Long Division

$$
\begin{array}{r}
x^2 + 4x + 3 \\
(x-1) \overline{) x^3 + 3x^2 - x - 3} \\
-\underline{(x^3 - x^2)} \\
4x^2 - x \\
-\underline{(4x^2 - 4x)} \\
3x - 3 \\
-\underline{(3x - 3)} \\
0
\end{array}
$$

Remainder 0

Synthetic Division

$$\boxed{1} \quad \begin{array}{ccccc} & 1 & 3 & -1 & -3 \\ & & 1 & 4 & 3 \\ \hline & 1 & 4 & 3 & 0 \end{array}$$

Answer

$$x^2 + 4x + 3 + \frac{0}{x-1}$$

Division of Polynomials – Ex. 01
Long Division

Use long division to evaluate: $\dfrac{x^3}{x^2+1}$

Note: All terms must be represented. (add fillers)

$$\begin{array}{r} x \\ (x^2+0x+1) \overline{\smash{\big)}\, x^3 + 0x^2 + 0x + 0} \\ -\underline{(x^3 + 0x^2 + 1x\,)} \\ -x \end{array}$$

Remainder (↑ pointing to $-x$)

Answer: $\dfrac{x^3}{x^2+1} = x - \dfrac{x}{x^2+1}$

Division of Polynomials – Ex. 02
Long Division & Synthetic Division

Use <u>long division</u> and <u>synthetic division</u> to evaluate:

$$\frac{x^2 + x^3 - 2x - 5}{x - 3}$$ Note: Put terms in proper order.

Solution

$$
\begin{array}{r}
x^2 + 4x + 10 \\
(x-3) \overline{\smash{)}\, x^3 + x^2 - 2x - 5} \\
-(x^3 - 3x^2) \\
\hline
4x^2 - 2x \\
-(4x^2 - 12x) \\
\hline
10x - 5 \\
-(10x - 30) \\
\hline
25
\end{array}
$$

Remainder → 25

Put the zero in the box.

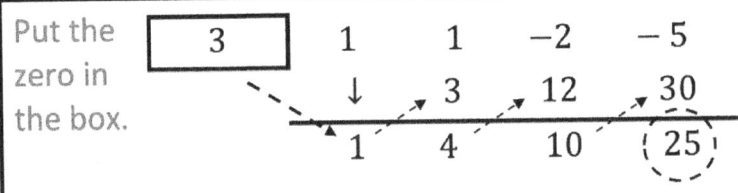

3	1	1	−2	−5
	↓	3	12	30
	1	4	10	(25)

Answer: $\dfrac{x^2 + x^3 - 2x - 5}{x - 3} = x^2 + 4x + 10 + \dfrac{25}{x - 3}$

Division of Polynomials – Ex. 03
Long Division & Synthetic Division

Use <u>long</u> and <u>synthetic</u> <u>division</u> to evaluate.	$\dfrac{2x^4 - x^3 + 4x^2 - 14x + 6}{2x - 1}$

Solution

$$\begin{array}{r} x^3 + 2x - 6 \\ (2x-1) \overline{\smash{\big)}\, 2x^4 - x^3 + 4x^2 - 14x + 6} \\ \underline{-\ (2x^4 - x^3)} \\ 0 + 4x^2 - 14x \\ \underline{-(4x^2 - 2x)} \\ -12x + 6 \\ \underline{-(-12x + 6)} \\ 0 \end{array}$$

$$\dfrac{2x^4 - x^3 + 4x^2 - 14x + 6}{2x - 1} = \dfrac{x^4 - \tfrac{1}{2}x^3 + 2x^2 - 7x + 3}{x - \tfrac{1}{2}}$$

$\boxed{\tfrac{1}{2}}$	1	$-\tfrac{1}{2}$	2	-7	3
	↓	$\tfrac{1}{2}$	0	1	-3
	1	0	2	-6	0

$$\dfrac{2x^4 - x^3 + 4x^2 - 14x + 6}{2x - 1} = x^3 + 2x - 6$$

Factoring Polynomials

Factoring Binomials: $a^2 - b^2 = (a-b)(a+b)$
The Difference of Two Squares

Example: Factor the binomial. $x^2 - 9$	
Rewrite as $a^2 - b^2$	$(x)^2 - (3)^2$
$x^2 - 9 = (x-3)(x+3)$	

Example: Factor the binomial. $4x^2 - 25$	
Rewrite as $a^2 - b^2$	$(2x)^2 - (5)^2$
$4x^2 - 25 = (2x-5)(2x+5)$	

Example: Factor the binomial. $32x^2 - 18$	
Factor then rewrite as $a^2 - b^2$	$2(16x^2 - 9)$ $2((4x)^2 - (3)^2)$
$32x^2 - 18 = 2(4x-3)(4x+3)$	

Factoring The Sum of Two Cubes

$$a^3 + b^3 = (a+b)(a^2 - ab + b^2)$$

Example: Factor the binomial. $8x^3 + 125$

Rewrite as $a^3 + b^3$	$(2x)^3 + (5)^3$

$$8x^3 + 125 = (2x+5)(4x^2 - 10x + 25)$$

Factoring The Difference of Two Cubes

$$a^3 - b^3 = (a-b)(a^2 + ab + b^2)$$

Example: Factor the binomial. $27x^3 - 8$

Rewrite as $a^3 - b^3$	$(3x)^3 - (2)^3$

$$27x^3 - 8 = (3x-2)(9x^2 + 6x + 4)$$

	Factoring Polynomials – Useful Formulas
Perfect Square Trinomials	$(a + b)^2 = a^2 + 2ab + b^2$ $(a - b)^2 = a^2 - 2ab + b^2$
Difference of 2 Squares	$a^2 - b^2 = (a - b)(a + b)$
Sum and Diff. of 2 Cubes	$a^3 + b^3 = (a + b)(a^2 - ab + b^2)$ $a^3 - b^3 = (a - b)(a^2 + ab + b^2)$
Distributive Property	$(a + b)(c + d + e)$ $= ac + ad + ae + bc + bd + be$
Quadratic Formula	If: $\quad ax^2 + bx + c = 0$ Then: $\quad x = \dfrac{-b \pm \sqrt{b^2 - 4ac}}{2a}$
Binomial Expansion	$(a + b)^n = \sum_{k=0}^{n} \binom{n}{k} a^{n-k} \cdot b^k$
Combination	$\binom{n}{k} = \dfrac{n!}{k! \cdot (n-k)!}$
Pascal's Triangle	1 1 1 1 2 1 1 3 3 1 1 4 6 4 1 1 5 10 10 5 1

Factoring Trinomials: $ax^2 + bx + c$
With $a = 1$ All Positive Terms

The goal of factoring trinomials is to rewrite it as the product of two linear factors, in the form:
$$(x - p)(x - q)$$

Example: Factor the trinomial. $x^2 + 6x + 8$

Start with this form. Note: $x \cdot x = x^2$	$(x \qquad)(x \qquad)$
Find two numbers: • $2 \cdot 4 = 8$ • $2 + 4 = 6$	$(x \quad 2)(x \quad 4)$ $(x + 2)(x + 4)$
Check Middle Term: • Multiply the <u>inner</u> & <u>outer</u> numbers. • When added, it should equal the middle term $(6x)$	$x^2 + 6x + 8$ $(x + 2)(x + 4)$ $2x$ $+$ $4x$ $=$ $6x$

$$x^2 + 6x + 8 = (x + 2)(x + 4)$$

Factoring Trinomials: $ax^2 + bx + c$ **With** $a = 1$ — **Negative** b	
Example: Factor the trinomial. $x^2 - 6x + 8$	
Start with this form. Note: $x \cdot x = x^2$	$(x \quad)(x \quad)$
Find two numbers: • $(-2)(-4) = 8$ • $-2 - 4 = -6$	$(x - 2)(x - 4)$
Check Middle Term: • Multiply the <u>inner</u> & <u>outer</u> numbers. • When added, it should equal the middle term $(-6x)$	$x^2 - 6x + 8$ $(x - 2)(x - 4)$ $-2x$ $+$ $-4x$ $=$ $-6x$
$x^2 - 6x + 8 = (x - 2)(x - 4)$	

Factoring Trinomials: $ax^2 + bx + c$
With $a = 1$ **Negative** c

Example: Factor the trinomial. $x^2 + 3x - 10$	
Start with this form. Note: $x \cdot x = x^2$	$(x \quad\;\;)(x \quad\;\;)$
Find two numbers: • $(-2)(5) = -10$ • $-2 + 5 = 3$	$(x - 2)(x + 5)$
Check Middle Term: • Multiply the <u>inner</u> & <u>outer</u> numbers. • When added, it should equal the middle term $(+3x)$	$x^2 + 3x - 10$ $(x - 2)(x + 5)$ $-\;2x$ $+$ $+\;5x$ $=$ $+\;3x$
$x^2 + 3x - 10 = (x - 2)(x + 5)$	

Factoring Trinomials: $ax^2 + bx + c$ With $a = 1$ Negative b and c	
Example: Factor the trinomial. $x^2 - 3x - 10$	
Start with this form. Note: $x \cdot x = x^2$	$(x \quad)(x \quad)$
Find two numbers: • $(-2)(5) = -10$ • $2 - 5 = -3$	$(x + 2)(x - 5)$
Check Middle Term: • Multiply the <u>inner</u> & <u>outer</u> numbers. • When added, it should equal the middle term $(-3x)$	$x^2 - 3x - 10$ $(x + 2)(x - 5)$ $+2x$ $+$ $-5x$ $=$ $-3x$
$x^2 - 3x - 10 = (x + 2)(x - 5)$	

Factoring Trinomials: $ax^2 + bx + c$ With $a \neq 1$	
Example: Factor the trinomial. $2x^2 + 8x + 6$	
Note: Common factor Factor out the 2	$2(x^2 + 4x + 3)$
Start with this form. Note: $x \cdot x = x^2$	$2(x\ \ \ )(x\ \ \ )$
Find two numbers: • $(1)(3) = 3$	$2(x + 1)(x + 3)$
Check Middle Term: • Multiply the <u>inner</u> & <u>outer</u> numbers. • When added, it should equal the middle term $(+4x)$	$x^2 + 4x + 3$ $(x + 1)(x + 3)$ $+\ 1x$ $+$ $+\ 3x$ $=$ $+\ 4x$
$2x^2 + 8x + 6 = 2(x + 1)(x + 3)$	

Factoring Trinomials: $ax^2 + bx + c$
With $a \neq 1$ **Negative** c

Example: Factor the trinomial. $2x^2 + x - 15$	
Start with this form. Note: $2x \cdot x = 2x^2$	$(2x \quad)(x \quad)$
Find two numbers: • $(-5)(3) = -15$	$(2x - 5)(x + 3)$
Check Middle Term: • Multiply the <u>inner</u> & <u>outer</u> numbers. • When added, it should equal the middle term $(+1x)$	$x^2 + x - 15$ $(2x - 5)(x + 3)$ $\quad - 5x$ $\quad +$ $\quad + 6x$ $\quad =$ $\quad + x$
$2x^2 + x - 15 = (2x - 5)(x + 3)$	

| Factoring Trinomials: $ax^2 + bx + c$ |
| With $a \neq 1$ Negative b & c |

Example: Factor the trinomial. $3x^2 - 2x - 8$

Start with this form. Note: $3x \cdot x = 3x^2$	$(3x \quad)(x \quad)$
Find two numbers: • $(4)(-2) = -8$	$(3x + 4)(x - 2)$
Check Middle Term: • Multiply the <u>inner</u> & <u>outer</u> numbers. • When added, it should equal the middle term $(-2x)$	$x^2 - 2x - 8$ $(3x + 4)(x - 2)$ $+ 4x$ $+$ $- 6x$ $=$ $- 2x$

$$3x^2 - 2x - 8 = (3x + 4)(x - 2)$$

Why Factor Polynomials

Why Factor Polynomials?

Often, y is a function of x. So, we say: $y = f(x)$
And often, the function is a polynomial.

If $y = f(x)$ and the function is fully factored.
Then, it will be in the form: $y = (x - p)(x - q)$

When the function is fully factored, it is easy to find the values of x that will make the function $= 0$
In the above example, $y = 0$ if $x = p$ or q

Zero Product Principle

If $a \cdot b = 0$ Then, $a = 0$ or $b = 0$

Example: Given $y = f(x) = x(x - 2)(x + 5)$
Find the values of x that make the function $= 0$

Answer: $x = \{0, 2, -5\}$

Conclusion: If $y = f(x)$ and $f(x)$ is fully factored. Then, it is easy to find the zeros of the function by using the Zero Product Principle.

Completing The Square

Equations With Perfect Squares Are Easy to Solve

Example #1	$x^2 = 25$ $\sqrt{x^2} = \pm\sqrt{25}$ $x = \pm 5$
Check the solutions	$(5)^2 = 25$ OK $(-5)^2 = 25$ OK
Conclusion	Equations with perfect squares are easy to solve. Just take the square root of both sides. Add the $\pm$ sign.

Example #2	$(x-3)^2 = 16$ $\sqrt{(x-3)^2} = \pm\sqrt{16}$ $x - 3 = \pm 4$ $x = 3 \pm 4 = 7, -1$
Conclusion	Equations with perfect squares are easy to solve. Remember to add $\pm$ sign.

Completing the Square

If a 2nd degree equation $ax^2 + bx + c = n$
We can rearrange it to have a perfect square, making it easier to solve. $(x + c)^2 = m$

The technique is explained with an example, below.

Example: Given $x^2 - 6x - 62 = 50$
Solve for x, by completing the square.

Separate terms with x from the numbers.	$x^2 - 6x = 62 + 50$ $x^2 - 6x = 112$
Take the coefficient of x, divide by 2, Then square it. Add this to both sides of eqn.	$x^2 - 6x + \left(\frac{6}{2}\right)^2 = 112 + \left(\frac{6}{2}\right)^2$ $x^2 - 6x + (3)^2 = 112 + (3)^2$ $(x-3)^2 = 121$
Take the square root of both sides. Remember the $\pm$ sign. Solve it.	$\sqrt{(x-3)^2} = \pm\sqrt{121}$ $(x-3) = \pm 11$ $x = 3 \pm 11$ $x = 14, -8$

Completing the Square – Example 1

Example: Given $x^2 + 20x - 5 = -65$
Solve for x, by completing the square.

Separate terms with x from the numbers.	$x^2 - 20x = -65 + 5$ $x^2 - 20x = -60$
Divide coefficient of x by 2, then square it. Add to both sides.	$x^2 - 20x + \left(\frac{20}{2}\right)^2 = -60 + \left(\frac{20}{2}\right)^2$ $x^2 - 6x + (10)^2 = -60 + (10)^2$ $(x - 10)^2 = 40$
Take the square root of both sides. Remember the $\pm$ sign Solve it.	$\sqrt{(x-10)^2} = \pm\sqrt{40}$ $(x - 10) = \pm 2\sqrt{10}$ $x \approx 10 \pm 6.3$ $x \approx 16.3, \ 3.7$

Completing the Square – Example 2

Example: Given $2x^2 - x + 7 = 10$
Solve for x, by completing the square.

Coeff. of x^2 must be one. Divide both sides by 2.	$x^2 - \frac{1}{2}x + \frac{7}{2} = \frac{10}{2}$
Separate terms with x from the numbers.	$x^2 - \frac{1}{2}x = \frac{3}{2}$
Divide coefficient of x by 2, then square it. Add to both sides.	$x^2 - \frac{1}{2}x + \left(\frac{1}{4}\right)^2 = \frac{3}{2} + \left(\frac{1}{4}\right)^2$ $x^2 - \frac{1}{2}x + \left(\frac{1}{4}\right)^2 = \frac{3}{2} + \left(\frac{1}{4}\right)^2$ $\left(x - \frac{1}{4}\right)^2 = \frac{24}{16} + \frac{1}{16} = \frac{25}{16}$
Take the square root of both sides. Remember the $\pm$ sign. Solve it.	$\left(x - \frac{1}{4}\right) = \pm\sqrt{\frac{25}{16}}$ $x = \frac{1}{4} \pm \frac{5}{4}$ $x = \frac{6}{4}, -\frac{4}{4} = \frac{3}{2}, -1$

Quadratic Formula

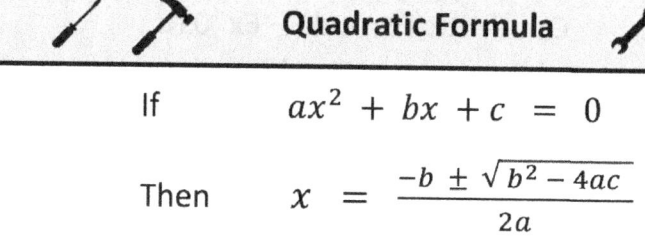

Quadratic Formula

If $ax^2 + bx + c = 0$

Then $x = \dfrac{-b \pm \sqrt{b^2 - 4ac}}{2a}$

The **Quadratic Formula** is one of the most important tools in your math tool-box. It is used in many math topics involving Quadratic Equations (2nd degree equations). **Memorize it!**

$a \neq 0$ for two reasons:
- If $a = 0$ then there is no x^2 term so it's a 1st degree equation.
- If $a = 0$ then the denominator, in the formula, is zero. Division by zero is undefined.

A little name confusion…
- "Quad" means four, but "Quadratic" comes from the Latin word "quadrum," meaning "to make square." But, don't worry about all of this!
- **Just remember, the Quadratic Formula is used to solve 2nd degree equations.**

Quadratic Formula – Ex. 01a
Solve Quadratic Equation (Solution #1)

Given: $x^2 + 2x - 8 = 0$

Solve the equation for x. In other words, find the value(s) of x that makes the equation true.

Solution #1: Use the Quadratic Formula

Identify the coefficients.	$a = 1, \quad b = 2, \quad c = -8$
Use the Quadratic Formula	$x = \dfrac{-b \pm \sqrt{b^2 - 4ac}}{2a}$ $x = \dfrac{-2 \pm \sqrt{(2)^2 - 4(1)(-8)}}{2(1)}$
Solve for x	$x = \dfrac{-2 \pm \sqrt{4 + 32}}{2} = \dfrac{-2 \pm \sqrt{36}}{2}$ $x = \dfrac{-2 \pm 6}{2} = \dfrac{-2}{2} \pm \dfrac{6}{2}$ $x = -1 \pm 3$ $x = 2, -4$

\multicolumn{2}{c}{**Quadratic Formula – Ex. 01b**}
\multicolumn{2}{c}{**Solve Quadratic Equation (Solution #2)**}

Given: $x^2 + 2x - 8 = 0$
Solve the equation for x. In other words, find the value(s) of x that makes the equation true.
Solution #2: Use factoring.

Factor the trinomial	$x^2 + 2x - 8 = 0$ $(x - 2)(x + 4) = 0$
Use the zero product principle. If $a \cdot b = 0$ Then $a = 0$ or $b = 0$	If either factor is zero, the left side is equal to zero. By observation … $x = 2, -4$
Conclusions	When solving a quadratic equation, if possible, factor it and just use the zero product principle. Just use the quadratic formula if you can't factor it!

Quadratic Formula – Ex. 02
Solve Quadratic Equation

Given: $6x^2 - 7x - 5 = 0$

Solve the equation, using the Quadratic Formula. Then, use your answer to factor the trinomial.

Identify the coefficients.	$a = 6, \quad b = -7, \quad c = -5$
Use the Quadratic Formula	$x = \dfrac{-b \pm \sqrt{b^2 - 4ac}}{2a}$ $x = \dfrac{7 \pm \sqrt{(-7)^2 - 4(6)(-5)}}{2(6)}$
Solve for x	$x = \dfrac{7 \pm \sqrt{49 + 120}}{12} = \dfrac{7 \pm \sqrt{169}}{12}$ $x = \dfrac{7 \pm 13}{12} = \dfrac{20}{12}, \dfrac{-6}{12}$ $x = \dfrac{5}{3}, -\dfrac{1}{2}$
Factor the trinomial	$6x^2 - 7x - 5 = 0$ $\left(x - \dfrac{5}{3}\right) \cdot \left(x + \dfrac{1}{2}\right) = 0$

Quadratic Formula – Ex. 03a
Solve Quadratic Equation (Solution #1)

Given: $x^2 - 9 = 0$

Solve the equation, using the Quadratic Formula. Then, solve it again, by factoring.

Solution #1: Use the Quadratic Formula

Note: This is a 2nd degree eqn.	$x^2 - 9 = 0$ $x^2 + 0x - 9 = 0$
Identify the coefficients.	$a = 1, \quad b = 0, \quad c = -9$
Use the Quadratic Formula	$x = \dfrac{-b \pm \sqrt{b^2 - 4ac}}{2a}$ $x = \dfrac{0 \pm \sqrt{0 - 4(1)(-9)}}{2(1)}$
Solve for x	$x = \dfrac{\pm\sqrt{36}}{2} = \dfrac{\pm 6}{2} = \pm 3$ $x = 3, -3$

Quadratic Formula – Ex. 03b
Solve Quadratic Equation (Solution #2)

Given: $x^2 - 9 = 0$
Solve the equation, using the Quadratic Formula.
Then, solve it again, by factoring.

Solution #2: Use factoring.

Factor the trinomial Note: Difference of two squares	$x^2 - 9 = 0$ $x^2 - 3^2 = 0$ $(x - 3)(x + 3) = 0$
Use the zero product principle. If $a \cdot b = 0$ Then $a = 0$ or $b = 0$	If either factor is zero, the left side is equal to zero. By observation … $x = 3, -3$
Conclusion	When solving a quadratic equation, if possible, factor it and just use the zero product principle.

Quadratic Formula – Ex. 04a
Solve Quadratic Equation (Solution #1)

Given: $x^2 - 7 = 0$
Solve the equation, using the Quadratic Formula.
Then, solve it again, by factoring.

Solution #1: Use the Quadratic Formula

Note: This is a 2nd degree eqn.	$x^2 - 7 = 0$ $x^2 + 0x - 7 = 0$	
Identify the coefficients.	$a = 1, \quad b = 0, \quad c = -7$	
Use the Quadratic Formula	$x = \dfrac{-b \pm \sqrt{b^2 - 4ac}}{2a}$ $x = \dfrac{0 \pm \sqrt{0 - 4(1)(-7)}}{2(1)}$	
Solve for x	$x = \dfrac{\pm \sqrt{28}}{2}$	$= \dfrac{\pm \sqrt{4 \cdot 7}}{2}$
	$x = \dfrac{\pm 2\sqrt{7}}{2}$	$= \pm \sqrt{7}$
	$x = \sqrt{7}, \ -\sqrt{7}$	

Quadratic Formula – Ex. 04b
Solve Quadratic Equation (Solution #2)

Given: $x^2 - 7 = 0$

Solve the equation, using the Quadratic Formula. Then, solve it again, by factoring.

Solution #2: Use factoring.

Factor the trinomial Note: Difference of two squares	$x^2 - 7 = 0$ $x^2 - (\sqrt{7})^2 = 0$ $(x - \sqrt{7})(x + \sqrt{7}) = 0$
Use the zero product principle. If $a \cdot b = 0$ Then $a = 0$ or $b = 0$	If either factor is zero, the left side is equal to zero. By observation ... $x = \sqrt{7}, -\sqrt{7}$
Conclusion	When solving a quadratic equation, if possible, factor it and just use the zero product principle.

Quadratic Formula – Ex. 05
Solve Quadratic Equation

Given: $4x^2 + 9 = 0$
Solve the equation, using the Quadratic Formula.
Then, use your answer to factor the trinomial.

Note: Can't factor using the difference of two squares because it's the sum of two squares!

Identify the coefficients.	$a = 4, \quad b = 0, \quad c = 9$
Use the Quadratic Formula	$x = \dfrac{-b \pm \sqrt{b^2 - 4ac}}{2a}$ $x = \dfrac{0 \pm \sqrt{0 - 4(4)(9)}}{2(4)}$
Solve for x	$x = \dfrac{\pm\sqrt{0 - 144}}{8} = \dfrac{\pm\sqrt{-1}\sqrt{144}}{8}$ $x = \dfrac{\pm 12i}{8} = \dfrac{\pm 3i}{2} = \dfrac{3i}{2}, \; -\dfrac{3i}{2}$
Factor the trinomial	$\left(x - \dfrac{3i}{2}\right) \cdot \left(x + \dfrac{3i}{2}\right) = 0$ $\dfrac{1}{2}(2x - 3i) \cdot \dfrac{1}{2}(2x + 3i) = 0$ $\dfrac{1}{4}(2x - 3i)(2x + 3i) = 0$

Function Composition

Function Composition	
\multicolumn{2}{l}{Function composition is an operation involving two functions (g and h) to produce a 3rd function (f). $$f = g \circ h$$ The composition of two functions, g and h has several notations, as shown below:}
$g \circ h$	Pronounced "g of h" or "g circle h" Here, output from h is input to g
$g(h(x))$	Pronounced "g of h of x" Here, x is input to h The output from $h(x)$ is the input to g

---

In general, $g(h(x)) \neq h(g(x))$

But if

$g(h(x)) = h(g(x)) = x$

then they are inverse functions.

Function Composition -- Example Set 1

Given: $g(x) = 3x$ and $h(x) = \sqrt{x}$
Find the following.

$g(h(4))$	$= g(\sqrt{4}) = g(2) = 3(2) = 6$
$h(g(4))$	$= h(3(4)) = h(12) = \sqrt{12} = 2\sqrt{3}$
$g(h(9))$	$= g(\sqrt{9}) = g(3) = 3(3) = 9$
$h(g(9))$	$= h(3(9)) = h(27) = \sqrt{27} = 3\sqrt{3}$
$g(h(x))$	$= g(\sqrt{x}) = 3\sqrt{x}$
$h(g(x))$	$= h(3x) = \sqrt{3x}$
$h \circ g$	$= h(g(x)) = h(3x) = \sqrt{3x}$

Function Composition -- Example Set 2

Given graphs of g and h, find the indicated values.

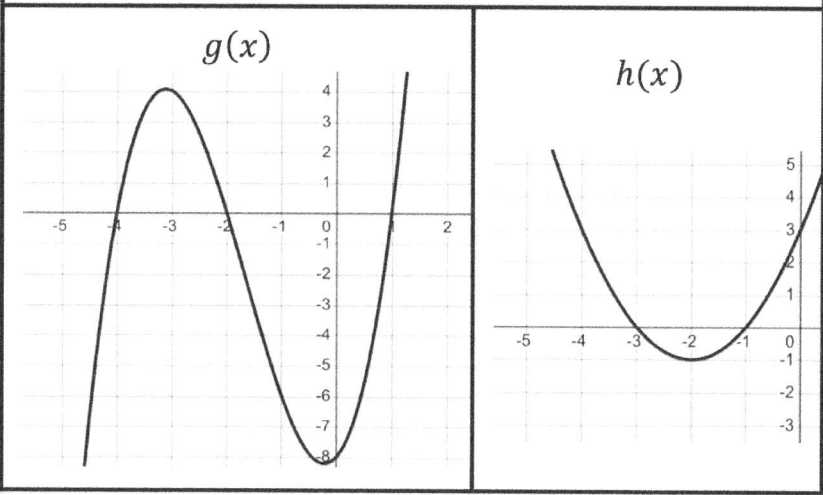

$g(h(-3))$	$= g(0) = -8$
$h(g(-4))$	$= h(0) = 3$
$g(h(-2))$	$= g(-1) = -6$
$g(h(-0.5))$	$= g(1) = 0$

Function Composition -- Example Set 3

Given values of g and h in the table below, find the indicated values.

$x \rightarrow$	0	1	2	3	4	5	6
$g(x)$	0	2	4	6	8	10	12
$h(x)$	5	3	1	0	4	6	2

$g(h(0))$	$= g(5) = 10$
$h(g(0))$	$= h(0) = 5$
$g(h(1))$	$= g(3) = 6$
$h(g(1))$	$= h(2) = 1$
$g(h(2))$	$= g(1) = 2$
$h(g(2))$	$= h(4) = 4$
$g(h(3))$	$= g(0) = 0$
$h(g(3))$	$= h(6) = 2$

Inverse Functions

Inverse of a Function

Let $f(x)$ and $g(x)$ be two **one-to-one** functions.

If $\quad f \circ g \ = \ g \circ f \ = \ x$

Then $\quad f(x)$ and $g(x)$ are inverses of each other.

Notation:
$\quad f^{-1}(x) \ = \ $ The inverse of $f(x)$
$\quad g^{-1}(x) \ = \ $ The inverse of $g(x)$

One-to-One Functions

A function is **one-to-one** if there is no horizontal line that intersects its graph more than once.

In other words ... A function is **one-to-one** if it passes the horizontal line test.

$y = x^3$ Is one-to-one	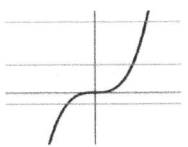
$y = x^2$ Is NOT one-to-one	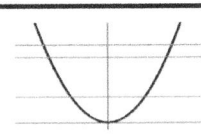

How to Find the Inverse of a Function
Basically, we just switch the $x's$ and $y's$
The simple example below, demonstrates the process.

Inverse Function – Example 0	
Given: $f(x) = 2x - 5$ Find: $f^{-1}(x)$	
Rewrite function as $y = f(x)$	$y = 2x - 5$
Switch the $x's$ and $y's$	$x = 2y - 5$
Solve for y	$x = 2y - 5$
$x + 5 = 2y$
$y = \dfrac{x+5}{2}$
Rename it to $f^{-1}(x)$

We can't call both functions " y " | $f^{-1}(x) = \dfrac{x+5}{2}$ |

How to Determine if Two Functions Are Inverses of Each Other

There are two ways to determine if two functions are inverses of each other – Algebraically and Graphically.

Given: $f = 2x - 5$ and $g = \dfrac{x+5}{2}$

Algebraic Test $f \circ g = x$ $g \circ f = x$	$f \circ g = f(g(x))$ $= g\left(\dfrac{x+5}{2}\right)$ $= 2\left(\dfrac{x+5}{2}\right) - 5 = x$
	$g \circ f = g(f(x))$ $= g(2x - 5)$ $= \dfrac{(2x-5) + 5}{2} = x$
Graphical Test Graphs of the two functions should be reflections over the $y = x$ axis	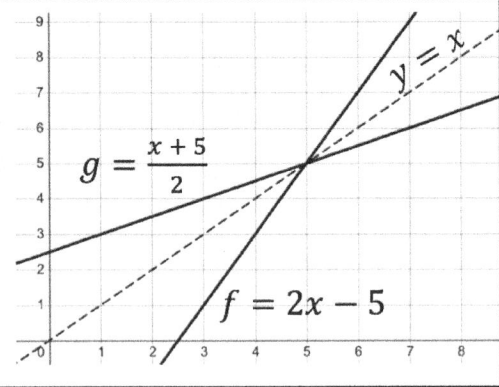

Inverse of a Function – Example Set 1

Given: $f(x) = x^3 + 7$	$y = x^3 + 7$
Switch x & y Solve for y	$x = y^3 + 7$ $y^3 = x - 7$ $y = \sqrt[3]{x - 7}$
Inverse of $f(x)$	$f^{-1}(x) = \sqrt[3]{x - 7}$

Given: $f(x) = \dfrac{x+3}{5}$	$y = \dfrac{x+3}{5}$
Switch x & y Solve for y	$x = \dfrac{y+3}{5}$ $5x = y + 3$ $y = 5x - 3$
Inverse of $f(x)$	$f^{-1}(x) = 5x - 3$

Given: $f(x) = \sqrt[3]{x+9}$	$y = \sqrt[3]{x+9}$
Switch x & y Solve for y	$x = \sqrt[3]{y+9}$ $x^3 = y + 9$ $y = x^3 - 9$
Inverse of $f(x)$	$f^{-1}(x) = x^3 - 9$

Inverse of a Function – Example 2

Given:	$f(x) = x^2 + 3$; $x \geq 0$
Graph it	*(graph of $f(x) = x^2 + 3$ for $x \geq 0$)*
Domain & Range In interval Notation	$D: [0, \infty)$ $R: [3, \infty)$
Write an equation for $f^{-1}(x)$	$y = x^2 + 3$; $x \geq 0$ $x = y^2 + 3$; $y \geq 0$ $y = \sqrt{x - 3}$ $f^{-1}(x) = \sqrt{x - 3}$
Graph $y = f(x)$ And $y = f^{-1}(x)$ On the same coordinate system.	*(graph showing both f and f^{-1} with line $y = x$)*

Note: Domains and Ranges are opposite (switched) For a function and its' inverse.

Transformations

Transformations
Transformations change a given function. Consider the function: $f(x) = x^2$ 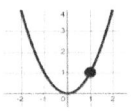

Horizontal Transformations (Very Nice)		
$y = f(x) + a$	V. Shift Up	
$y = a * f(x)$	V. Stretch	
$y = -f(x)$	V. Rotation	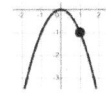

Horizontal Transformations (Horrible)		
$y = f(x + a)$	H. Shift Left	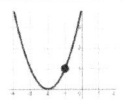
$y = f(a * x)$	H. Compression	 
$y = f(-x)$	H. Rotation	

Transformations – Ex. 1a

Given: $y = (x+3)^2 - 5$
Find: Parent function, transformations, and graph.

Parent Function $y = x^2$	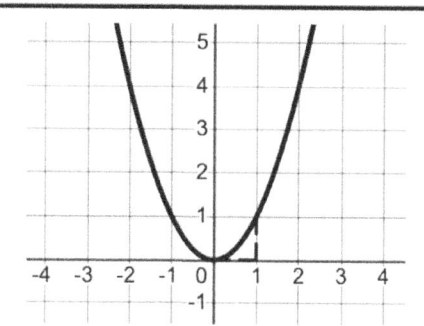	
Vertical and Horizontal Transformations	• Horizontal shift left by 3 • Vertical shift down by 5 • No stretching or compression	
Graph $y = (x+3)^2 - 5$	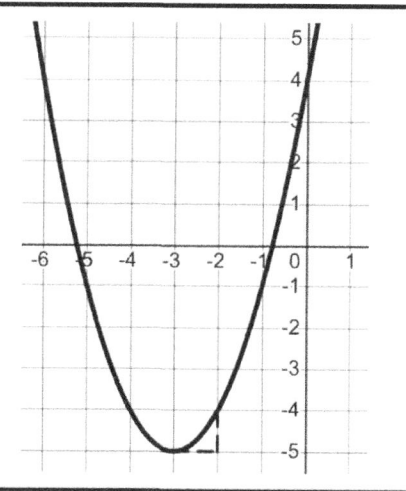	

Transformations – Ex. 1b (With T table)

Given: $y = (x+3)^2 - 5$
Find: Parent function, transformations, and graph.

Parent Function $y = x^2$	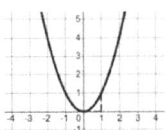
Vertical and Horizontal Transformations	• Horizontal shift left by 3 • Vertical shift down by 5 • No stretching or compression

Use transformations to create extended "T" table.

$x - 3$	x	y	$y - 5$
-3	0	0	-5
-2	1	1	-4
-1	2	4	-1

Use points from table to help graph

$y = (x+3)^2 - 5$

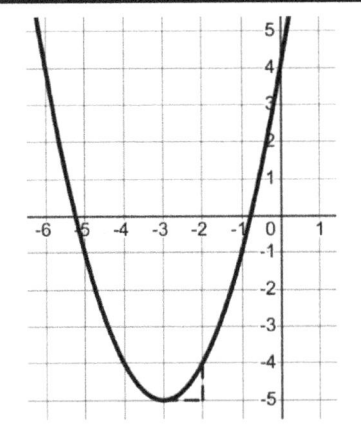

Transformations – Ex. 2a

Given: $y = -\sqrt{2x+6}$
Find: Parent function, transformations, and graph.

Parent Function $y = \sqrt{x}$	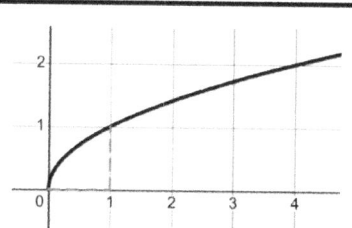
Rewrite eqn. so coefficient of x is 1	$y = -\sqrt{2x+6}$ $y = -\sqrt{2(x+3)}$
Vertical and Horizontal Transformations	• H. Shift left by 3 • H. Compression by ½ • V. Rotation over x axis
Graph $y = -\sqrt{2x+6}$	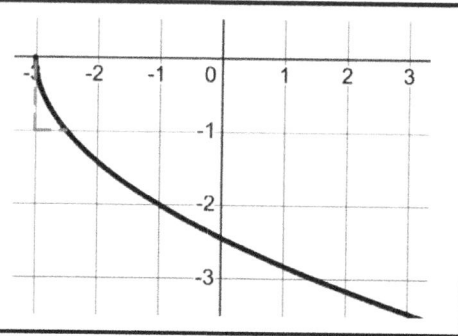

Transformations – Ex. 2b (With T table)

Given: $y = -\sqrt{2x+6}$
Find: Parent function, transformations, and graph.

Parent Function $y = \sqrt{x}$	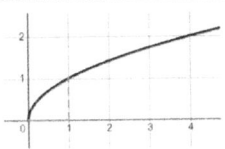
Rewrite eqn. so coeff. of x is 1	$y = -\sqrt{2x+6}$ $y = -\sqrt{2(x+3)}$
Vertical and Horizontal Transformations	• H. Compression by ½ • H. Shift left by 3 • V. Rotation over x axis
Use transformations to create extended T-table	$\frac{1}{2}x - 3$ x y $-y$ -3 0 0 0 -2.5 1 1 -1 -1 2 4 -2
Use points from T table to help graph $y = -\sqrt{2x+6}$	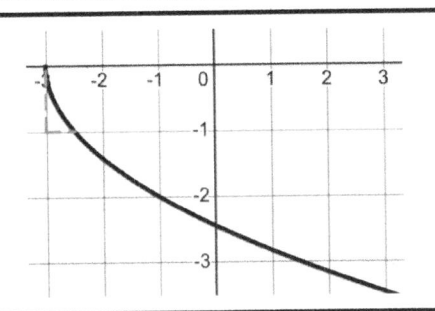

Transformations – Ex. 3a

Given: $y = -3|x-2|$
Find: Parent function, transformations, and graph.

Parent Function $y = |x|$	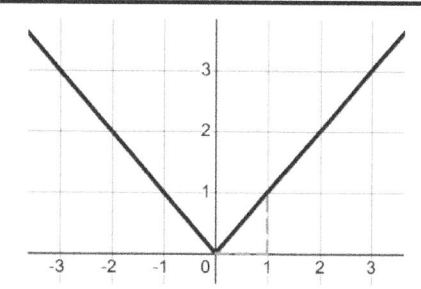	
Vertical and Horizontal Transformations	• H. Shift right by 2 • V. Stretch by 3 • V. Rotation over x axis	
Graph $y = -3|x-2|$	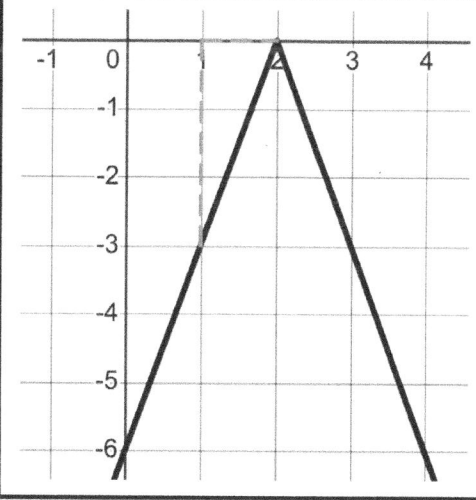	

	Transformations – Ex. 3b (With T table)
Given: $y = -3|x - 2|$ Find: Parent function, transformations, and graph.	
Parent Function $y = |x|$	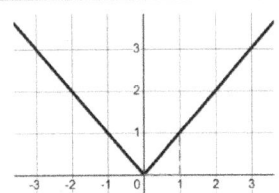
Vertical and Horizontal Transformations	• H. Shift right by 2 • V. Stretch by 3 • V. Rotation over x axis
Use transformations to create extended T-table	<table><tr><td>$x+2$</td><td>x</td><td>y</td><td>$-3y$</td></tr><tr><td>2</td><td>0</td><td>0</td><td>0</td></tr><tr><td>3</td><td>1</td><td>1</td><td>-3</td></tr><tr><td>1</td><td>-1</td><td>1</td><td>-3</td></tr></table>
Use points from T-table to help graph $y = -3|x - 2|$	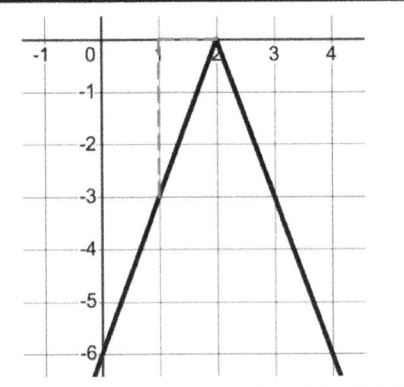

Transformations – Ex. 4

Given: The graph. Find: • Parent function, • Transformations • Equation	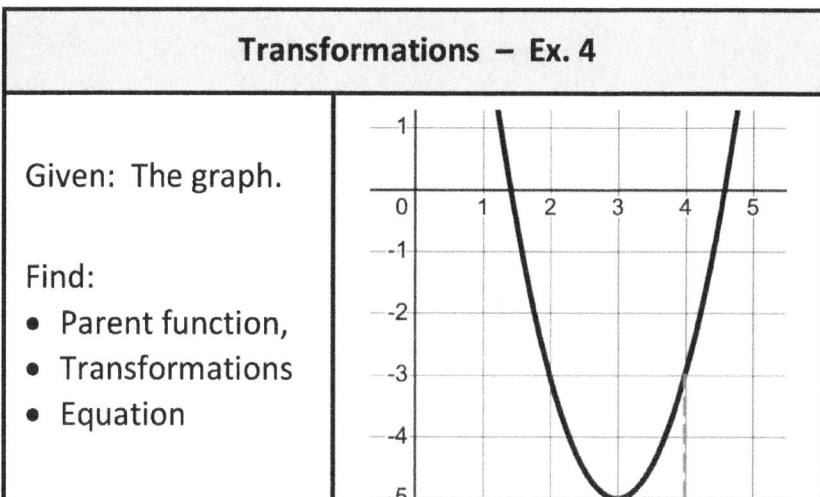

Parent Function $y = x^2$	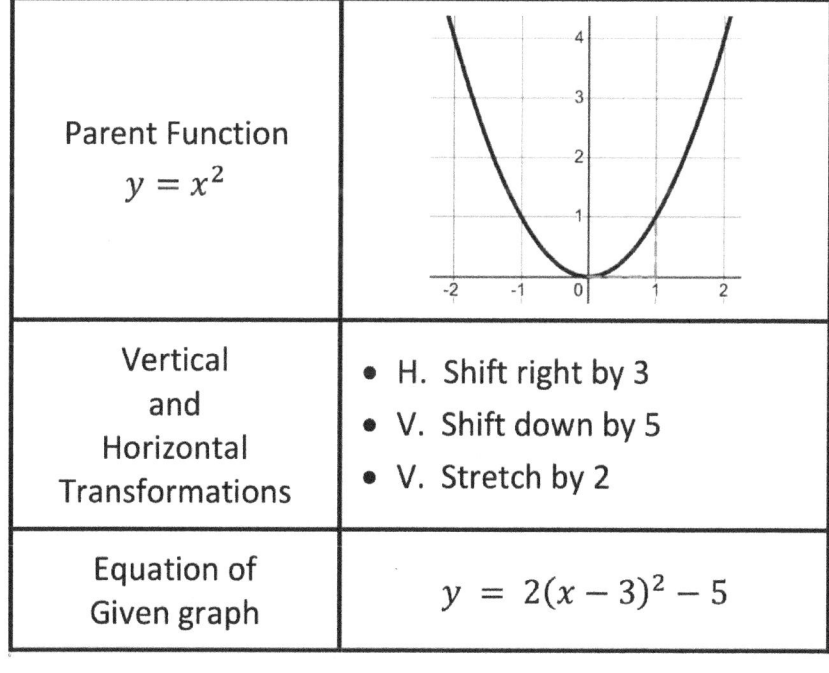
Vertical and Horizontal Transformations	• H. Shift right by 3 • V. Shift down by 5 • V. Stretch by 2
Equation of Given graph	$y = 2(x-3)^2 - 5$

Transformations – Ex. 5

Given: The graph. Find: • Parent function, • Transformations • Equation	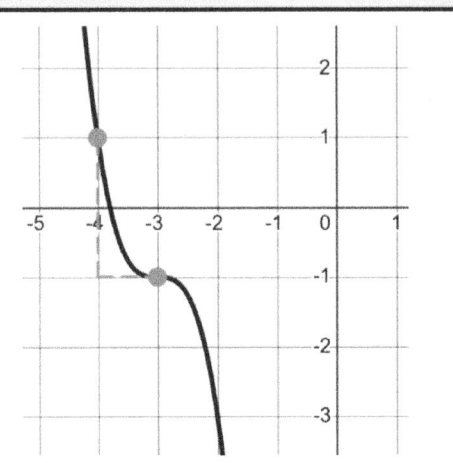
Parent Function $y = x^3$	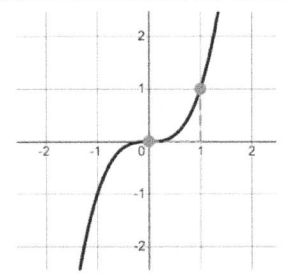
Vertical and Horizontal Transformations	• H. Shift left by 3 • V. Shift down by 1 • V. Stretch by 2 • V. Rotation over x axis
Equation of Given graph	$y = -2(x+3)^3 - 1$

Transformations – Ex. 6

Given: The graph. Find: • Parent function, • Transformations • Equation	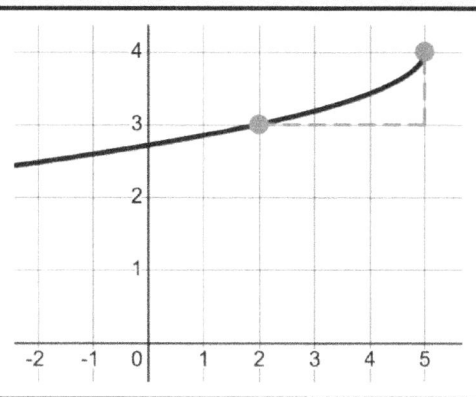
Parent Function $y = \sqrt{x}$	
Vertical and Horizontal Transformations	• H. Shift right by 5 • H. Stretch left by 3 • H. Rotation over y axis • V. Rotation over x axis • V. Shift up by 4 (do last)
Equation of Given graph	$y = -\sqrt{-\frac{1}{3}(x-5)} + 4$

	Absolute Value Transformations
	Taking the absolute value of the **output** is different than taking the absolute value of the **input.**

$y = \|f(x)\|$	Absolute value of the output. If you take the absolute value of the output, then all output is positive. All " y " values are positive.

$y = f(\|x\|)$	Absolute value of the input. Here, the " x " values can still be negative or positive. They just act like positive x input. Graph has y axis symmetry.

Transformations – Ex. 1
Absolute Value of the Output: $y = |f(x)|$

Given : A set of (x, y) points Answer the questions below.	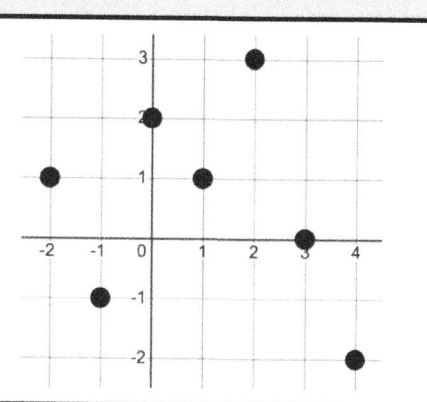

Find the domain & range of original set of (x, y) points.	$D: \{-2, -1, 0, 1, 2, 3, 4\}$ $R: \{-2, -1, 0, 1, 2, 3\}$
Find the domain & range of $(x, |y|)$ points.	$D: \{-2, -1, 0, 1, 2, 3, 4\}$ $R: \{0, 1, 2, 3\}$
Plot the set of points $(x, |y|)$ Note: All y values are positive	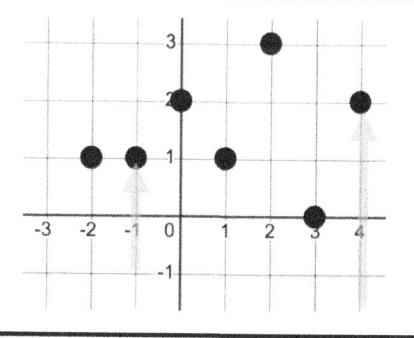

Transformations – Ex. 2
Absolute Value of the Output: $y = |f(x)|$

Given:
$y = (x-1)^2 - 2$

Answer the questions below.

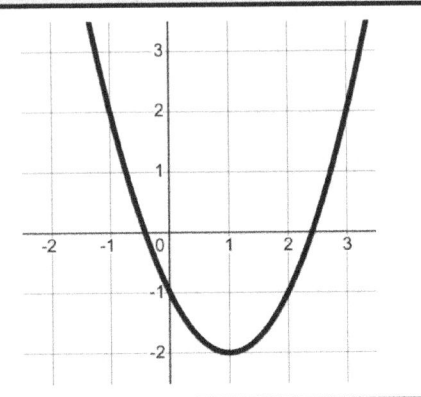

Find the domain & range of original function.	$D: (-\infty, \infty)$ $R: [-2, \infty)$
Find the domain & range of $y = |(x-1)^2 - 2|$	$D: (-\infty, \infty)$ $R: [0, \infty)$
Plot the function $y = |(x-1)^2 - 2|$ Note: All y values are positive	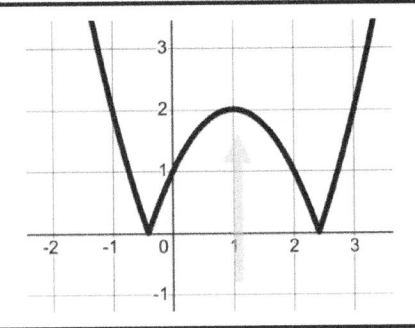

Transformations – Ex. 3
Absolute Value of the Input: $y = f(|x|)$

Given: A set of (x, y) points Answer the questions below.	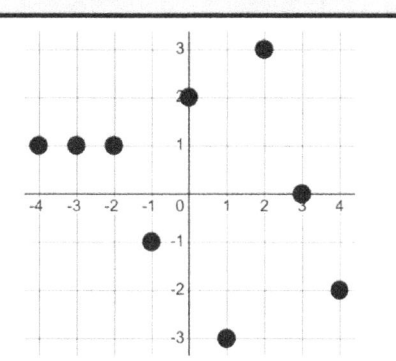

Find the domain & range of original set of (x, y) points.	$D: \{-4, -3, -2, -1, 0, 1, 2, 3, 4\}$ $R: \{-3, -2, -1, 0, 1, 2, 3\}$
Find the domain & range of $(|x|, y)$ points.	$D: \{-4, -3, -2, -1, 0, 1, 2, 3, 4\}$ $R: \{-3, -2, 0, 1, 3\}$
Plot the set of points $(x, |y|)$ Note: y-axis symmetry	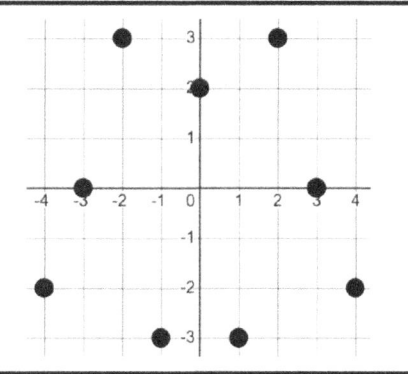

Transformations – Ex. 4
Absolute Value of the Input: $y = f(|x|)$

Given: $y = (x-1)^2 - 2$ Answer the questions below.	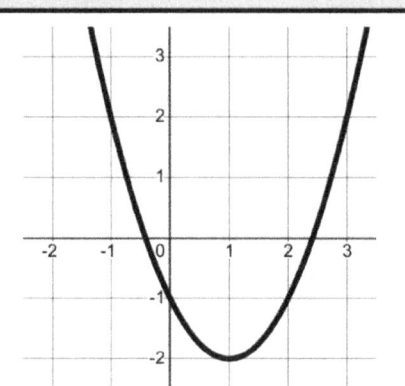

Find the domain & range of original function.	D: $(-\infty, \infty)$ R: $[-2, \infty)$
Find the domain & range of $y = (|x|-1)^2 - 2$	D: $(-\infty, \infty)$ R: $[-2, \infty)$
Plot the function $y = (|x|-1)^2 - 2$ Note: y-axis symmetry	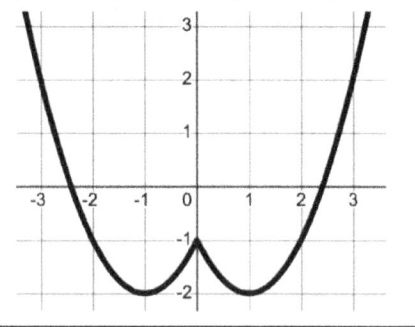

Graphing Quadratic Functions

Graphing Quadratic Functions
Quadratic Functions may be in one of three forms.

Quadratic Equations -- Vertex		
Form	Equation	Vertex (h, k)
Standard Form	$f(x) = ax^2 + bx + c$	$h = -\frac{b}{2a}$ $k = f(h)$
Vertex Form	$f(x) = a(x - h)^2 + k$	See eqn.
Factored Form	$f(x) = a(x - z_1)(x - z_2)$	$h = \frac{z_1 + z_2}{2}$ $k = f(h)$

Quadratic Equations – Far End Behavior		
$a =$ Positive	Happy Parabola (opens up)	☺
$a =$ Negative	Sad Parabola (opens down)	☹

Graphing Quadratic Functions
Set the quadratic function $= 0$ to find the zeros (or x-intercepts). Use the Quadratic Formula to solve it.

Quadratic Formula		
If $\quad ax^2 + bx + c = 0$		
Then $\quad x = \dfrac{-b \pm \sqrt{b^2 - 4ac}}{2a}$		
Discriminant $= b^2 - 4ac$		
$b^2 - 4ac$	Zeros	Examples
0	1 Real	
Positive	2 Real	
Negative	0 Real	

Graphing Quadratic Functions – Ex. 1

Given: $f(x) = x^2 + 2x - 15$.   (2nd degree eqn.)
Find: Vertex, Zeros, Graph, Domain, Range.

Note: 2 points needed to graph 1st degree equation.
       3 points needed to graph 2nd degree equation.

Vertex	$h = -\dfrac{b}{2a} = -\dfrac{2}{2(1)} = -1$ $k = f(-1) = (-1)^2 + 2(-1) - 15$ $k = 1 - 2 - 15 = -16$ Vertex: $(h, k) = (-1, -16)$
Zeros	$x^2 + 2x - 15 = 0$ $(x - 3)(x + 5) = 0 \quad \rightarrow \quad x = 3, -5$ Zeros: $(3, 0), (-5, 0)$
Notes	3 Points: $(-1, -16)$, $(3, 0)$, $(-5, 0)$ Positive a: Opens up (Happy Parabola)
Graph	$D: (-\infty, \infty)$ $R: [-16, \infty)$

	Graphing Quadratic Functions – Ex. 2
	Given: $f(x) = x^2 - 2x - 5$ (2nd degree eqn.) Find: Vertex, Zeros, Graph, Domain, Range.
Vertex	$h = -\dfrac{b}{2a} = \dfrac{2}{2(1)} = 1$ $k = f(1) = (1)^2 - 2(1) - 5 = -6$ Vertex: $(h, k) = (1, -6)$
Zeros	$x^2 - 2x - 5 = 0$ Use Quadratic Formula $x = \dfrac{-b \pm \sqrt{b^2 - 4ac}}{2a}$ $x = \dfrac{2 \pm \sqrt{4 - 4(1)(-5)}}{2(1)} = \dfrac{2 \pm \sqrt{24}}{2} = \dfrac{2 \pm 2\sqrt{6}}{2}$ $x = 1 \pm \sqrt{6}$ Zeros: $(1 + \sqrt{6}, 0), (1 - \sqrt{6}, 0)$
Notes	3 Points: Two zeros and vertex (see above) Positive a: Opens up (Happy Parabola)
Graph	$D: (-\infty, \infty)$ $R: [-6, \infty)$

Graphing Quadratic Functions – Ex. 3

Given: $f(x) = -x^2 + 10x - 25$ (2nd degree eqn.)
Find: Vertex, Zeros, Graph, Domain, Range.

Vertex	$h = -\dfrac{b}{2a} = \dfrac{-10}{2(-1)} = 5$ $k = f(5) = -(5)^2 + 10(5) - 25 = 0$ Vertex: $(h, k) = (5, 0)$
Zeros	$-x^2 + 10x - 25 = 0$ Use Quad. Formula $x = \dfrac{-b \pm \sqrt{b^2 - 4ac}}{2a}$ $x = \dfrac{-10 \pm \sqrt{100 - 4(-1)(-25)}}{2(-1)} = \dfrac{-10 \pm \sqrt{0}}{-2}$ $x = 5$ → One Zero: $(5, 0)$
Notes	Vertex: $(5, 0)$, Negative a: Opens down Find more points: $f(0) = -25$ $f(1) = -1 + 10 - 25 = -16$ 3 Points: $(5, 0), (0, -25), (1, -16)$
Graph	$D: (-\infty, \infty)$ $R: (-\infty, 0\,]$

Graphing Quadratic Functions – Ex. 4

Given the graph.

Find the equation.

Use Vertex Form With $(h, k) = (3, -4)$	$y = a(x - h)^2 + k$ $y = a(x - 3)^2 - 4$
Use another point $(x, y) = (1, 4)$ to find " a "	$y = a(x - 3)^2 - 4$ $(1, 4) \rightarrow 4 = a(1 - 3)^2 - 4$ $4 = a(-2)^2 - 4$ $4 = a(4) - 4$ $8 = a(4)$ $2 = a$
Equation	$y = 2(x - 3)^2 - 4$

Graphing Quadratic Functions – Ex. 5

Given graph. Find equation.	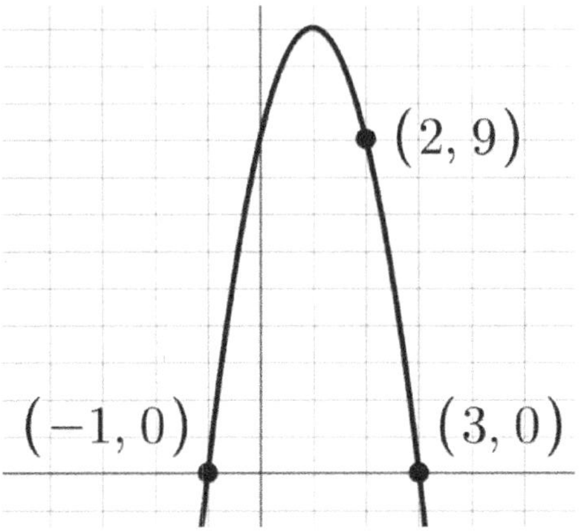 Parabola through $(-1, 0)$, $(3, 0)$, and $(2, 9)$

Use Factored Form With $z_1 = -1$ $z_2 = 3$	$y = a(x - z_1)(x - z_2)$ $y = a(x + 1)(x - 3)$
Use another point $(x, y) = (2, 9)$ to find " a "	$y = a(x + 1)(x - 3)$ $(2, 9) \rightarrow 9 = a(2 + 1)(2 - 3)$ $9 = a(3)(-1)$ $9 = a(-3)$ $-3 = a$ (Sad Parabola)
Equation	$y = -3(x + 1)(x - 3)$
Note: Equation could be expanded to Standard Form	

Graphing Quadratic Functions (Maximum) – Ex. 6	
Given: $f(x) = -3(x+1)(x-3)$ Find: The maximum or minimum value.	

Maximum or Minimum?	Leading coefficient is negative so it is a "sad" parabola. It opens downward. So, it has a maximum and no minimum.
Maximum at vertex.	Maximum occurs at the vertex, on the axis of symmetry (AOS). AOS is between the two x-intercepts or between the two zeros (z_1 and z_2).
Find the zeros.	Since the given function is in factored form, the zeros are easy to find. $z_1 = -1 \qquad z_2 = 3$
Find the vertex.	Vertex at $x = \dfrac{z_1 + z_2}{2} = \dfrac{-1+3}{2} = 1$ $y = f(1) = -3(1+1)(1-3) = 12$ Vertex is at $(x, y) = (1, 12)$
Maximum	Maximum value of the function $= 12$

Graphing Quadratic Functions (Minimum) – Ex. 7

Given: $f(x) = 3x^2 + 9x - 5$

Find: The maximum or minimum value.

Maximum or Minimum?	Leading coefficient is positive so it is a "happy" parabola. Opens upward. So, it has a minimum but no max.
Find the vertex.	Minimum occurs at the vertex, where $x = -\dfrac{b}{2a} = -\dfrac{9}{2(3)} = -\dfrac{3}{2}$ $y = 3\left(-\dfrac{3}{2}\right)^2 + 9\left(-\dfrac{3}{2}\right) - 5 = -\dfrac{47}{4}$ Vertex at $(x, y) = \left(-\dfrac{3}{2}, -\dfrac{47}{4}\right)$
Minimum	Minimum value of the function $= -\dfrac{47}{4}$
Graph	(graph of parabola with vertex at $(-1.5, -11.75)$)

Graphing Quadratic Functions (Intercepts) – Ex. 8a

Given: $f(x) = x^2 + 3x - 4$

Find: All intercepts and vertex. Then, graph it.

x intercepts occur when $y = 0$	$x^2 + 3x - 4 = 0$ $(x - 1)(x + 4) = 0$ $x = 1, -4$
y intercept occurs when $x = 0$	$f(0) = 0^2 + 3(0) - 4 = .-4$ y-intercept at $(x, y) = (0, -4)$
Vertex	AOS between the x intercepts $h = \frac{1 + (-4)}{2} = -\frac{3}{2}$ $k = f\left(-\frac{3}{2}\right) = \left(-\frac{3}{2}\right)^2 + 3\left(-\frac{3}{2}\right) - 4$ $k = \frac{9}{4} - \frac{9}{2} - 4$ $k = \frac{9}{4} - \frac{18}{4} - \frac{16}{4} = -\frac{25}{4} = -6.25$ Vertex at $(h, k) = (-1.5, -6.25)$

Graphing Quadratic Functions (Intercepts) – Ex. 8b

Given: $f(x) = x^2 + 3x - 4$

Find: All intercepts and vertex. Then, graph it.

Previously Found	x intercepts: $x = 1, -4$ y intercepts: $y = -4$ Vertex: $(h, k) = (-1.5, -6.25)$
Graph	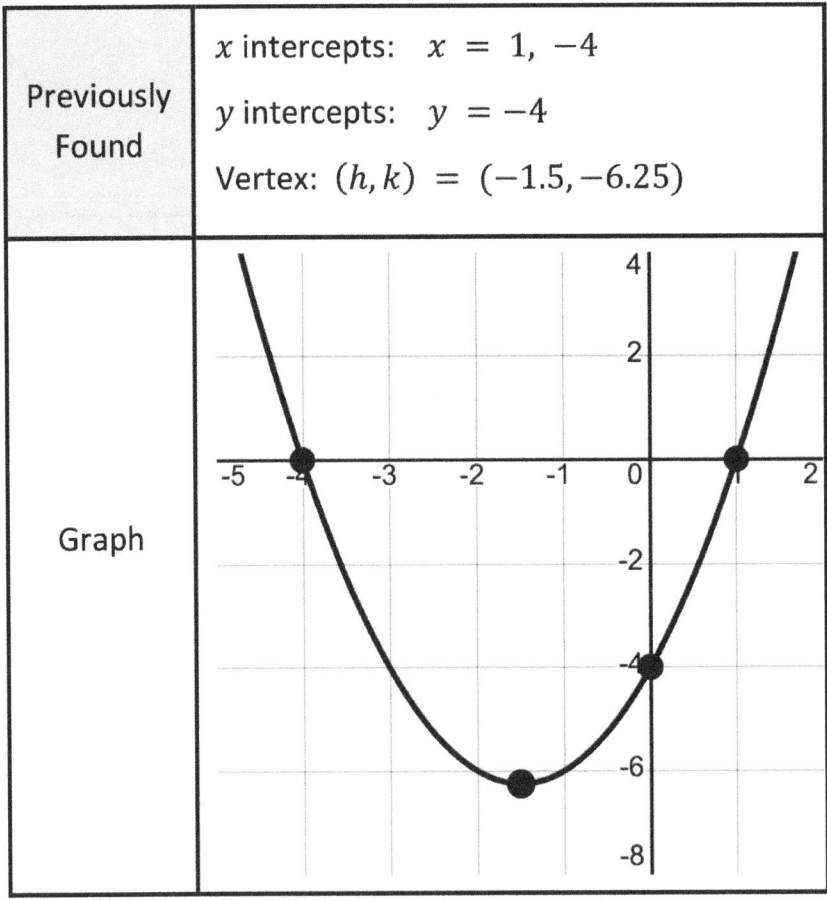

Graphing Polynomial Functions

Graphing Polynomial Functions

In the previous section, we graphed quadratic functions, which are 2nd degree functions.
In this section, we will graph polynomial functions, which are functions with degrees ≥ 1.

Polynomial functions in factored form:
$$y = a(x - z_1)^i (x - z_2)^j \ldots (x - z_n)^k$$
Degree = Sum of exponents = $i + j + \cdots + k$

Polynomial Functions – Far End Behavior

Far-end behavior depends on the degree		Similar to	Positive a	Negative a
	Odd Degree	1st Degree	↙↗	↖↘
	Even Degree	2nd Degree	↖↗	↙↘

Odd Degree
Positive "a"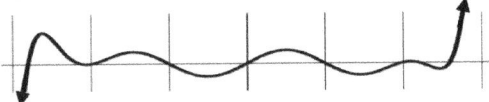

Turning Points	Number of times a polynomial function changes direction (or wiggles around). Number of turning points = Degree -1

Graphing Polynomial Functions (Continued)

The zeros (or roots) of a polynomial function are easy to find if the polynomial is in factored form:
$$y = a(x-z_1)^i(x-z_2)^j \ldots (x-z_n)^k$$
Degree = Sum of exponents = $i + j + \cdots + k$

Polynomial Functions – Behavior at Zeros

Behavior at zeros depends on multiplicity.	The multiplicity of a zero is the number of times it occurs. The above polynomial has the following zeros and multiplicities. • z_1 Multiplicity = i • z_2 Multiplicity = j • z_n Multiplicity = k	
	Odd Multiplicity	Polynomial goes through the zero.
	Even Multiplicity	Polynomial bounces at the zero.
Example Positive "a" Even Degree	Bounce Thru Thru Bounce	

Graphing Polynomial Functions -- Ex. 1

Given: $f(x) = \frac{1}{2}(x+1)^2(x-2)^3(x-4)$

Find: Degree, Zeros (x intercepts), Multiplicity, Far-End behavior, and y-intercept. Then sketch it.

Degree	$D = 2 + 3 + 1 = 6$ (even)
Zeros and Multiplicity	$-1, m2$ $2, m3$ $4, m1$ Bounce Thru. Thru.
Far-End Behavior	Positive leading coefficient and even degree. ↖ ↗
y-intercept	y-intercept occurs at $x = 0$ $f(0) = \frac{1}{2}(0+1)^2(0-2)^3(0-4)$ $f(0) = \frac{1}{2}(1)(-8)(-4) = 16$
Graph	

Graphing Polynomial Functions -- Ex. 2	
Given: $f(x) = -2x^4(x+3)^2$ Find: Degree, Zeros (x intercepts), Multiplicity, Far-End behavior, and y-intercept. Then sketch it.	
Degree	$D = 4 + 2 = 6$ (even)
Zeros and Multiplicity	$-0, m4$ $-3, m2$ Bounce Bounce
Far-End Behavior	Negative leading coefficient and even degree. ↙ ↘
y-intercept	y-intercept occurs at $x = 0$ $f(0) = -2(0)^4(0+3)^2$ $f(0) = -2(0)(3) = 0$ Point (0,0)
Graph	

Graphing Polynomial Functions -- Ex. 3

Given: $f(x) = 2x^2 + 9x - 5$

Find: Degree, Zeros (x intercepts), Multiplicity, Far-End behavior, and y-intercept. Then sketch it.

Factor it	$f(x) = (2x - 1)(x + 5)$
Degree	$D = 1 + 1 = 2$ (even)
Zeros and Multiplicity	$\frac{1}{2}$, $m1$ -5, $m1$ Thru Thru
Far-End Behavior	Positive leading coefficient and even degree. ↖ ↗
y-intercept	y-intercept occurs at $x = 0$ $f(0) = (-1)(5) = -5$ Point $(0, -5)$
Graph	

Graphing Polynomial Functions -- Ex. 4

Given: $f(x) = 3x^2 - 5x - 2$

Find: Degree, Zeros (x intercepts), Multiplicity, Far-End behavior, and y-intercept. Then sketch it.

Factor it	$f(x) = (3x + 1)(x - 2)$ $f(x) = 3\left(x + \frac{1}{3}\right)(x - 2)$
Degree	$D = 1 + 1 = 2$ (even)
Zeros and Multiplicity	$-\frac{1}{3}$, m1 2, m1 Thru Thru
Far-End Behavior	Positive leading coefficient and even degree. ↖ ↗
y-intercept	y-intercept occurs at $x = 0$ $f(0) = -2$ Point: $(0, -2)$
Graph	

Graphing Rational Functions

Graphing Rational Functions

Rational Functions are functions in the form:

$$R(x) = \frac{f(x)}{g(x)} \quad ; g(x) \neq 0$$

D_n, D_d = Degrees of numerator & denom.
a, b = Leading coeff. of numerator & denom.

Asymptote	#	Look for this	Form
Vertical (VA)	Can be many	Values that make denom. = 0	$x = c$
Horizontal (HA)	One at most	$D_n < D_d$	$y = 0$
		$D_n = D_d$	$y = \frac{a}{b}$
Slant (SA)	One at most	$D_n = D_d + 1$	$y = mx + b$
Hole ◯	Can be many	Zeros of factors that cancel.	$(c, 0)$

The function will never cross a vertical asymptote.

The function may cross horizontal & slant asymptotes. To check, evaluate: $R(x) = HA$ or $R(x) = SA$

Graphing Rational Functions – Ex. 1a

Given: $y = \dfrac{2(x-1)(x+2)(x-4)}{3(x-1)(x+3)(x-5)} = \dfrac{2(x+2)(x-4)}{3(x+3)(x-5)}$

Find: All asymptotes, holes, and zeros. Then graph.

Degrees	$D_n = 3 \qquad D_d = 3$
Vertical (VA)	$x = -3, \, 5$
Horizontal (HA)	$y = \dfrac{2}{3} \qquad D_n = D_d$
Slant (SA)	None.
Hole	Hole at $x = 1$ \qquad Point: $\left(1, \dfrac{3}{8}\right)$ $y = \dfrac{2(1+2)(1-4)}{3(1+3)(1-5)} = \dfrac{-18}{-48} = \dfrac{3}{8}$
x –intercepts	$y = 0 \;\rightarrow\;$ Numerator $= 0$ $x = -2, \, 4$
y –intercept	$x = 0 \;\rightarrow\; y = \dfrac{2(2)(-4)}{3(3)(-5)}$ $y = \dfrac{-16}{-45} \approx 0.35$
Does function cross HA ?	See next page.

Graphing Rational Functions – Ex. 1b

Given: $y = \dfrac{2(x-1)(x+2)(x-4)}{3(x-1)(x+3)(x-5)} = \dfrac{2(x+2)(x-4)}{3(x+3)(x-5)}$

Find: All asymptotes, holes, and zeros. Then graph.

Previously Found		
	VA	$x = -3, 5$
	HA	$y = \dfrac{2}{3}$
	Hole	Point: $\left(1, \dfrac{3}{8}\right)$
	x–int'cpt	$y = 0 \rightarrow x = -2, 4$
	y–int'cpt	$x = 0 \rightarrow y \approx 0.35$

Does function cross HA ?	$\dfrac{2(x+2)(x-4)}{3(x+3)(x-5)} = \dfrac{2}{3}$
	$\dfrac{(x+2)(x-4)}{(x+3)(x-5)} = 1$
	$(x+2)(x-4) = (x+3)(x-5)$
	$x^2 - 2x - 8 = x^2 - 2x - 15$
	$-8 = -15$
	NOT POSSIBLE $\rightarrow$ DOES NOT CROSS

Graphing Rational Functions – Ex. 1c

Given: $y = \dfrac{2\,(x-1)(x+2)(x-4)}{3\,(x-1)(x+3)(x-5)} = \dfrac{2\,(x+2)(x-4)}{3\,(x+3)(x-5)}$

Find: All asymptotes, holes, and zeros. Then graph.

Previously Found	VA	$x = -3, 5$
	HA	$y = \dfrac{2}{3}$ Does NOT cross
	Hole	Point: $\left(1, \dfrac{3}{8}\right)$
	x –int'cpt	$y = 0 \rightarrow x = -2, 4$
	y –int'cpt	$x = 0 \rightarrow y \approx 0.35$
Graph	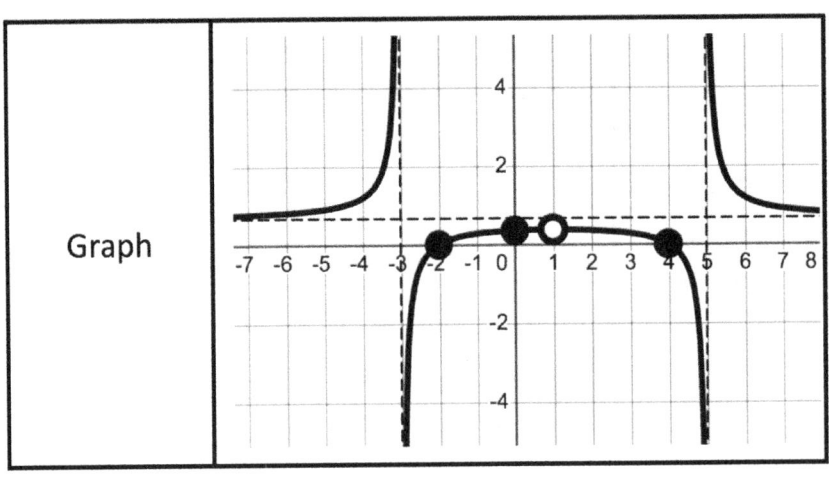	

Graphing Rational Functions – Ex. 2a	
Given: $y = \dfrac{6x^2 + 7}{x + 1}$ Find: All asymptotes, holes, and zeros. Then graph.	
Degrees	$D_n = 2 \quad D_d = 1$
Vertical (VA)	$x = -1$
Horizontal (HA)	None.
Slant (SA)	$D_n = D_d + 1 \rightarrow$ There is a SA Divide numerator by denominator and ignore the remainder. $\begin{array}{r} 6x \;-\; 6 \end{array}$ $x+1 \,\overline{\smash{\big)}\, 6x^2 + 0x + 7}$ $\underline{-\;(6x^2 - 6x\,)}$ $ 6x + 7$ $\underline{-(-6x - 6\,)}$ $ 13$ SA: $y = 6x - 6$
Hole	None.
Does function cross SA ?	See next page.

	Graphing Rational Functions – Ex. 2b	
Given: $y = \dfrac{6x^2 + 7}{x + 1}$ Find: All asymptotes, holes, and zeros. Then graph.		
Previously Found	VA	$x = -1$
	SA	$y = 6x - 6$
Does function cross SA ?	$\dfrac{6x^2 + 7}{x + 1} = 6x - 6$ $6x^2 + 7 = (6x - 6)(x + 1)$ $6x^2 + 7 = 6(x - 1)(x + 1)$ $6x^2 + 7 = 6(x^2 - 1)$ $6x^2 + 7 = 6x^2 - 6$ $7 = 6$ NOT POSSIBLE → Does not cross SA	
x –intercepts	$y = 0 \;\to\; \text{Numerator} = 0$ $6x^2 + 7 = 0$ $x = \pm\sqrt{-\dfrac{7}{6}}$ Imaginary number No x-intercepts	
y –intercept	$x = 0 \;\to\; y = \dfrac{0 + 7}{0 + 1} = 7$ Point: $(0, 7)$	

Graphing Rational Functions – Ex. 2c

Given: $y = \dfrac{6x^2 + 7}{x + 1}$

Find: All asymptotes, holes, and zeros. Then graph.

Previously Found	VA	$x = -1$
	SA	$y = 6x - 6$ Does not cross
	x Int'cpt	No x intercepts
	y Int'cpt	Point: $(0, 7)$

Graph	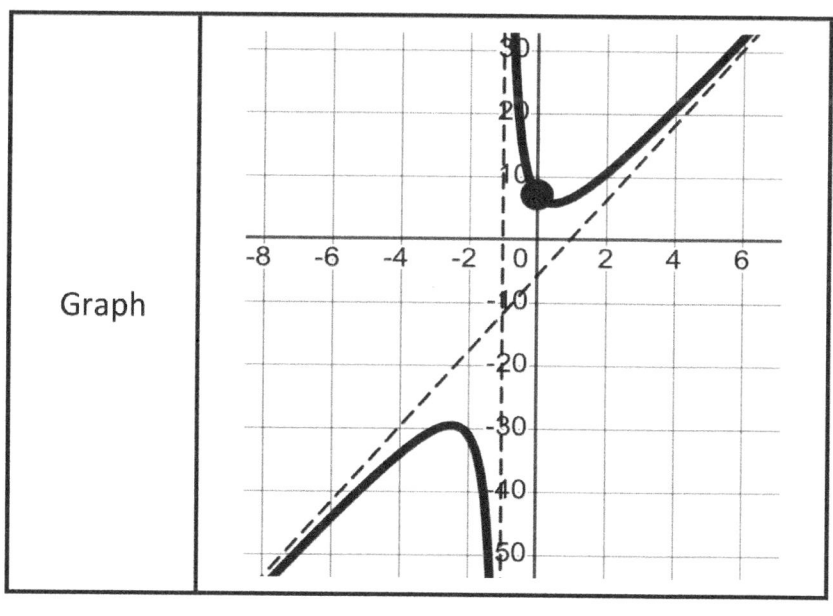

Graphing Rational Functions – Ex. 3a

Given: $y = \dfrac{1}{x}$

Find: All asymptotes, holes, and zeros. Then graph.

Degrees	$D_n = 0 \quad D_d = 1$
Vertical (VA)	$x = 0$ \quad Makes denom. $= 0$
Horizontal (HA)	$y = 0 \quad D_d > D_n$
Slant (SA)	None.
Holes	None.
Does function cross HA ?	$\dfrac{1}{x} = 0$ $1 = 0$ \quad NOT Possible Function does NOT cross HA
x-intercepts	$y = 0 \;\rightarrow\;$ Numerator $= 0$ NOT Possible. $\rightarrow$ No x-intercepts
y-intercept	$x = 0 \;\rightarrow\; y = \dfrac{1}{x} = \dfrac{1}{0}$ NOT Possible $\rightarrow$ No y-intercept
Graph	Next page...

Graphing Rational Functions – Ex. 3b

Given: $y = \dfrac{1}{x}$

Find: All asymptotes, holes, and zeros. Then graph.

Previously Found	VA	$x = 0$	Multiplicity $= 1$ Odd Multiplicity Function up on one side and down on other.
	HA	$y = 0$	Does not cross.
	x-int'cpt	None.	
	y-int'cpt	None.	

Graph	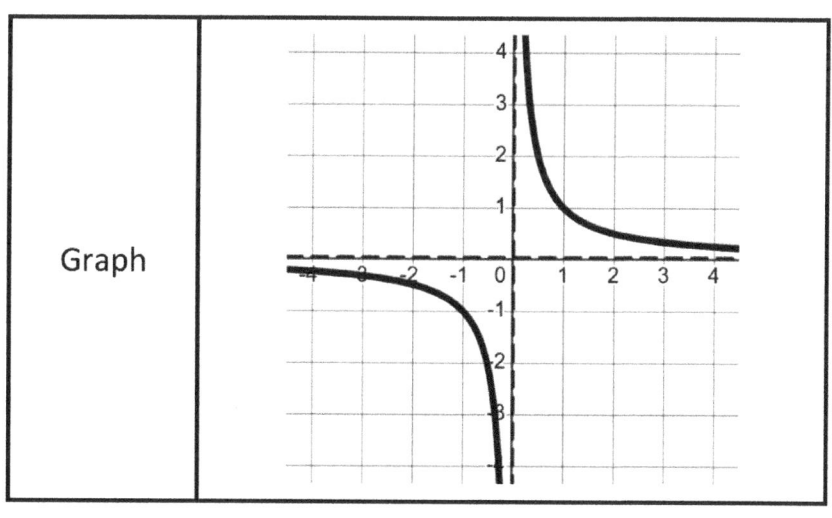

Graphing Rational Functions – Ex. 4a

Given: $y = \dfrac{1}{x^2}$

Find: All asymptotes, holes, and zeros. Then graph.

Degrees	$D_n = 0 \qquad D_d = 2$
Vertical (VA)	$x = 0$ \qquad Makes denom. $= 0$
Horizontal (HA)	$y = 0$ \qquad $D_d > D_n$
Slant (SA)	None.
Holes	None.
Does function cross HA ?	$\dfrac{1}{x^2} = 0$ $1 = 0$ \qquad NOT Possible Function does NOT cross HA
x-intercepts	$y = 0 \;\rightarrow\;$ Numerator $= 0$ NOT Possible. $\rightarrow$ No x-intercepts
y-intercept	$x = 0 \;\rightarrow\; y = \dfrac{1}{x} = \dfrac{1}{0}$ NOT Possible $\rightarrow$ No y-intercept
Graph	Next page…

Graphing Rational Functions – Ex. 4b

Given: $y = \dfrac{1}{x^2}$

Find: All asymptotes, holes, and zeros. Then graph.

Previously Found			
	VA	$x = 0$	Multiplicity = 2. Even Multiplicity. Function goes both up or both down near VA.
	HA	$y = 0$	Does not cross.
	x-int'cpt	None.	
	y-int'cpt	None.	

Graph

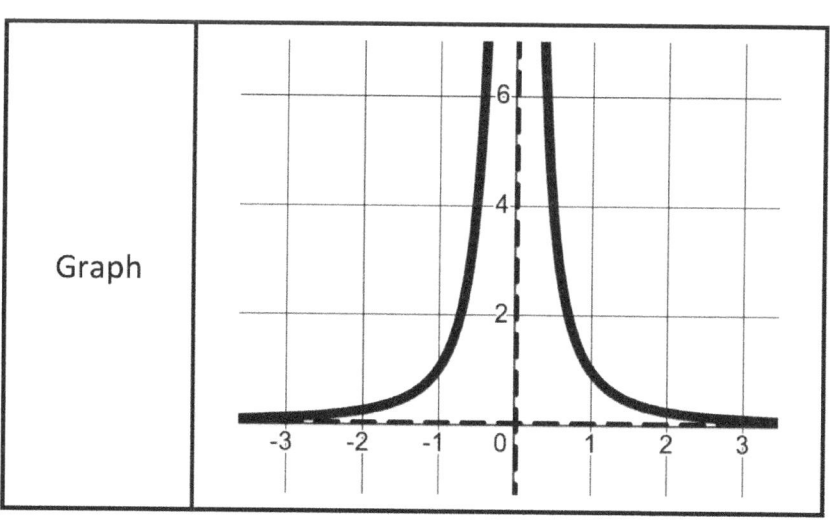

Graphing Rational Functions – Ex. 5a

Given: $f(x) = \dfrac{x^2 - 4}{x^2 - 4x} = \dfrac{(x-2)(x+2)}{x(x-4)}$

Find: All asymptotes, holes, and zeros. Then graph.

Degrees	$D_n = 2 \qquad D_d = 2$
Vertical (VA)	$x = 0 \quad$ and $\quad x = 4$
Horizontal (HA)	$y = \dfrac{1}{1} = 1$
Slant (SA)	None.
Hole	None.
Does function cross HA ?	$\dfrac{x^2 - 4}{x^2 - 4x} = 1$ $x^2 - 4 = x^2 - 4x$ $-4 = -4x$ $\dfrac{-4}{-4} = \dfrac{-4x}{-4}$ $1 = x \qquad$ This is possible. Function crosses HA at $x = 1$ $f(1) = \dfrac{(1-2)(1+2)}{1(1-4)} = \dfrac{-3}{-3} = 1$ Point: $(1, 1)$

Graphing Rational Functions – Ex. 5b

Given: $f(x) = \dfrac{x^2 - 4}{x^2 - 4x} = \dfrac{(x-2)(x+2)}{x(x-4)}$

Find: All asymptotes, holes, and zeros. Then graph.

Previously Found	VA	$x = 0$, $x = 4$
	HA	$y = 1$ Function crosses at point: $(1, 1)$
x-intercepts		$y = 0 \rightarrow$ Numerator $= 0$ $x = \pm 2$
y-intercept		$x = 0 \rightarrow y = \dfrac{0^2 - 4}{0^2 - 4(0)} = \dfrac{-4}{0}$ DNE No y-intercept
Graph		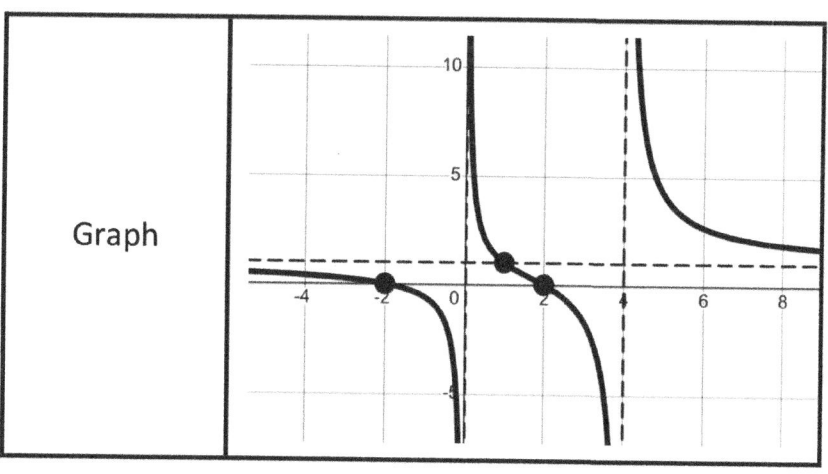

Graphing Rational Functions – Ex. 6a

Given: $f(x) = \dfrac{3x^2 - 4x}{x^2 + 6}$

Find: All asymptotes, holes, and zeros. Then graph.

Degrees	$D_n = 2$ $D_d = 2$
Vertical (VA)	None.
Horizontal (HA)	$y = \dfrac{3}{1} = 3$
Slant (SA)	None.
Hole	None.
Does function cross HA ?	$\dfrac{3x^2 - 4x}{x^2 + 6} = 3$ $3x^2 - 4x = 3x^2 + 18$ $-4x = 18$ $x = \dfrac{18}{-4} = -\dfrac{9}{2} = -4.5$ $x = -4.5$ This is possible. Function crosses HA at $x = -4.5$ Point: $(-4.5, 3)$

Graphing Rational Functions – Ex. 6b

Given: $f(x) = \dfrac{3x^2 - 4x}{x^2 + 6}$

Find: All asymptotes, holes, and zeros. Then graph.

Previously Found	VA	None.
	HA	$y = 3$ Function crosses HA at Point: $(-4.5, 3)$
x-intercepts		$y = 0 \rightarrow$ Numerator $= 0$ $x(3x - 4) = 0$ $x = 0, \dfrac{4}{3}$ Points: $(0,0)$, $\left(\dfrac{4}{3}, 0\right)$
y-intercept		$x = 0 \rightarrow y = \dfrac{0}{6} = 0$ Point: $(0, 0)$
Graph		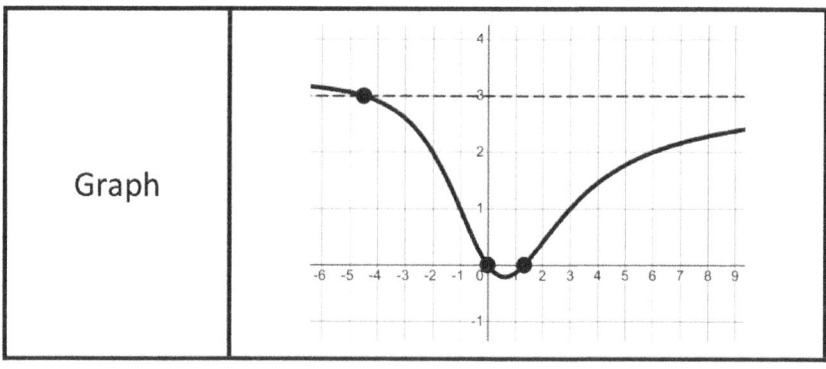

Working With Radicals

Rationalize the Denominator

Radicals in the denominator are often removed by **Rationalizing the Denominator**. Radicals in the numerator are okay.

Recall: $a^2 - b^2 = (a-b)(a+b)$

And: $\left[\sqrt{a}\right]^2 = a$

$y = \dfrac{1}{\sqrt{123}}$	$y = \dfrac{1}{\sqrt{123}}\ (1)$ $y = \dfrac{1}{\sqrt{123}}\left(\dfrac{\sqrt{123}}{\sqrt{123}}\right) = \dfrac{\sqrt{123}}{123}$
$y = \dfrac{1}{2+\sqrt{3}}$	$y = \dfrac{1}{2+\sqrt{3}}\ (1)$ $y = \dfrac{1}{2+\sqrt{3}}\left(\dfrac{2-\sqrt{3}}{2-\sqrt{3}}\right)$ $y = \dfrac{2-\sqrt{3}}{4-3} = \dfrac{2-\sqrt{3}}{1} = 2-\sqrt{3}$
$y = \dfrac{1}{6-\sqrt{7}}$	$y = \dfrac{1}{6-\sqrt{7}}\ (1)$ $y = \dfrac{1}{6-\sqrt{7}}\left(\dfrac{6+\sqrt{7}}{6+\sqrt{7}}\right)$ $y = \dfrac{6+\sqrt{7}}{36-7} = \dfrac{6+\sqrt{7}}{29}$

Working With Radicals

Recall, a radical is in the form: $\sqrt[n]{Radicand}$
Where: $n = $ index
Default: $n = 2$ Therefore: $\sqrt[2]{25} = \sqrt{25} = 5$

Even n	$y = \sqrt[n]{x}$ $D: [0, \infty)$ $R: [0, \infty)$	
Odd n	$y = \sqrt[n]{x}$ $D: (-\infty, \infty)$ $R: (-\infty, \infty)$	

When n is even: The radicand ≥ 0	Notes: $\sqrt{9} = 3$ $\sqrt{-9} = Undefined$ $\sqrt[3]{8} = 2$ $\sqrt[3]{-8} = -2$
When solving equations with radicals, isolate one radical and solve.	Note: $[\sqrt[n]{f(x)}]^n = f(x)$

	Working With Radicals -- Ex. 1

Given: $f(x) = \sqrt{3x + 12}$
Find: Domain and Range. Then, graph it.
Note: Using translations may help.

Domain & Range	$n =$ even Radicand ≥ 0 $3x + 12 \geq 0$ $x \geq -\frac{12}{3}$ $D: [-4, \infty)$	$f(x) = \sqrt{3x + 12}$ $f(-4) = 0$ $R: [0, \infty)$
Translations Parent: $y = \sqrt{x}$	$y = \sqrt{3x + 12}$ $y = \sqrt{3(x + 4)}$	H. Comp. by $\frac{1}{3}$ H. Shift left by 4 No Vert. Trans.
Graph		

Equations With Radicals -- Ex. 1

Solve for x: $3x - 2 = 5\sqrt{x}$
Check your answer(s) for extraneous solutions.

Isolate term with radical Then square both sides	$3x - 2 = 5\sqrt{x}$ $[3x - 2]^2 = \left[5\sqrt{x}\right]^2$ $(3x - 2)(3x - 2) = 25x$ $9x^2 - 12x + 4 = 25x$ $9x^2 - 37x + 4 = 0$ $x = \dfrac{37 \pm \sqrt{(-37)^2 - 4(9)(4)}}{2(9)}$ $x = \dfrac{37 \pm \sqrt{1225}}{18} = \dfrac{37 \pm 35}{18}$ $x = \dfrac{72}{18}, \dfrac{2}{18} = 4, \dfrac{1}{9}$	
Check answers	$x = 4$	$3(4) - 2 = 5\sqrt{4}$ $10 = 10$ TRUE
Use Original Equation	$x = \dfrac{1}{9}$	$3\left(\dfrac{1}{9}\right) - 2 = 5\sqrt{\dfrac{1}{9}}$ $-\dfrac{5}{3} = \dfrac{5}{3}$ FALSE
Solution	$x = 4$	

Equations With Radicals -- Ex. 3a

Solve for x: $\sqrt{2x+5} - 10 = 2\sqrt{2x} - 9$
Check your answer(s) for extraneous solutions.

Recall: $(a+b)^2 = a^2 + 2ab + b^2$
$\quad\quad\quad (a-b)^2 = a^2 - 2ab + b^2$

Isolate one radical term then square both sides.

Do this twice.

$$\sqrt{2x+5} = 2\sqrt{2x} + 1$$

$$\left[\sqrt{2x+5}\right]^2 = \left[2\sqrt{2x}+1\right]^2$$

$$2x + 5 = (2\sqrt{2x}+1)(2\sqrt{2x}+1)$$

$$2x + 5 = 4(2x) + 4\sqrt{2x} + 1$$

$$2x + 5 = 8x + 4\sqrt{2x} + 1$$

$$-6x + 4 = 4\sqrt{2x}$$

$$-3x + 2 = 2\sqrt{2x}$$

$$[2-3x]^2 = \left[2\sqrt{2x}\right]^2$$

$$4 - 12x + 9x^2 = 4(2x)$$

$$9x^2 - 20x + 4 = 0$$

$$x = \frac{20 \pm \sqrt{(-20)^2 - 4(9)(4)}}{2(9)} \quad \text{Quadratic Equation}$$

$$x = \frac{20 \pm \sqrt{256}}{18} = \frac{20 \pm 16}{18} = 2, \frac{2}{9}$$

Equations With Radicals -- Ex. 3b		
Solve for x: $\sqrt{2x+5} - 10 = 2\sqrt{2x} - 9$ Check your answer(s) for extraneous solutions. Recall: $(a+b)^2 = a^2 + 2ab + b^2$ $(a-b)^2 = a^2 - 2ab + b^2$		
Previously Found	$x = 2, \dfrac{2}{9}$	
Check answers Use Original Equation	$x = 2$	$\sqrt{2x+5} = 2\sqrt{2x} + 1$ $\sqrt{4+5} = 2\sqrt{4} + 1$ $3 = 4 + 1$ $3 = 5 \qquad\qquad$ FALSE
	$x = \dfrac{2}{9}$	$\sqrt{2x+5} = 2\sqrt{2x} + 1$ $\sqrt{\dfrac{4}{9} + \dfrac{45}{9}} = 2\sqrt{\dfrac{4}{9}} + 1$ $\sqrt{\dfrac{49}{9}} = 2\left(\dfrac{2}{3}\right) + 1$ $\dfrac{7}{3} = \dfrac{4}{3} + \dfrac{3}{3} \qquad$ TRUE
Solution	$x = \dfrac{2}{9}$	

Equations With Radicals -- Ex. 4a		
Solve for x: $\dfrac{2x}{\sqrt{x-1}} = x$		
Domain	$x - 1 > 0$ $x > 1$	Radicand must be ≥ 0 AND Denominator $\neq 0$ So, $x > 1$
Here, we can start by squaring both sides.	$\left[\dfrac{2x}{\sqrt{x-1}}\right]^2 = [x]^2$ $\dfrac{4x^2}{x-1} = x^2$ $4x^2 = x^2(x-1)$	
Don't divide by x. Factor instead.	You can not divide by x because x might $= 0$. Division by zero is undefined. $4x^2 = x^3 - x^2$ $0 = x^3 - 5x^2$ $x^2(x - 5) = 0$ $x = 0, 5$	
Note	$x = 0$ is not in the domain. If $x = 0$, denominator is imaginary. Therefore, $x = 5$	

Equations With Radicals -- Ex. 4b			
Solve for x: $\dfrac{2x}{\sqrt{x-1}} = x$			
Previously Found	Domain		$x > 1$
	One Solution in domain		$x = 5$
Check	$x = 5 \rightarrow$ $\dfrac{2x}{\sqrt{x-1}} = x$ $\dfrac{2(5)}{\sqrt{5-1}} = 5$ $\dfrac{10}{\sqrt{4}} = 5$ $\dfrac{10}{2} = 5$		TRUE

Equations With Radicals -- Ex. 5	
Solve for x: $\sqrt[3]{2x-5} + 7 = 4$	
Isolate one radical term then cube both sides.	$\sqrt[3]{2x-5} = -3$ $\left[\sqrt[3]{2x-5}\right]^3 = [-3]^3$ $2x - 5 = (-3)(-3)(-3)$ $2x - 5 = -27$ $2x = -22$ $x = -11$
No need to check for extraneous roots with cube roots (odd index) But check anyway!	$\sqrt[3]{2x-5} + 7 = 4$ $\sqrt[3]{2(-11)-5} + 7 = 4$ $\sqrt[3]{-22-5} + 7 = 4$ $\sqrt[3]{-27} + 7 = 4$ $(-3) + 7 = 4$ $4 = 4$ TRUE

The Inverse of $y = \sqrt{x}$ -- Ex. 6

Given: $f(x) = \sqrt{x}$
Find: Domain, range, and inverse of $f(x)$
And: Domain and range of $f^{-1}(x)$
Then: Sketch both $f(x)$ & $f^{-1}(x)$ on same graph.

$y = f(x)$	$y = \sqrt{x}$	D: $[0, \infty)$ R: $[0, \infty)$
Find the Inverse of $f(x)$	$y = \sqrt{x}$ $x = \sqrt{y}$ $x^2 = y$ $\rightarrow$	Original Equation Switch x and y $f^{-1}(x) = x^2$
Let $g = f^{-1}(x)$	$g = x^2$	D: $[0, \infty)$ R: $[0, \infty)$ Domain & Inverses are switched. They're the same, so not a big deal.
Graph $y = f(x)$ $g = f^{-1}(x)$		

	The Inverse of $y = \sqrt{x-5}$ -- Ex. 7	
Given:	$f(x) = \sqrt{x-5}$	
Find:	Domain, range, and inverse of $f(x)$	
And:	Domain and range of $f^{-1}(x)$	
Then:	Sketch both $f(x)$ & $f^{-1}(x)$ on same graph.	
$y = f(x)$	$y = \sqrt{x-5}$ D: $[5, \infty)$ R: $[5, \infty)$	
Find the Inverse of $f(x)$	$y = \sqrt{x-5}$ Original Equation $x = \sqrt{y-5}$ Switch x and y $x^2 = y - 5$ → $f^{-1}(x) = x^2 + 5$	
Let $g = f^{-1}(x)$	$g = x^2 + 5$ D: $[5, \infty)$ R: $[5, \infty)$ Domain & Inverses are switched. They're the same, so not a big deal.	
Graph $y = f(x)$ $g = f^{-1}(x)$	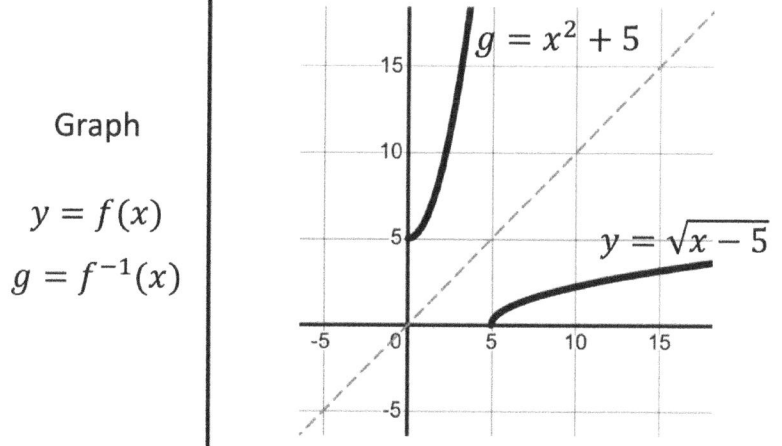	

The Inverse of $y = \sqrt[3]{x-5}$ -- Ex. 8

Given: $f(x) = \sqrt[3]{x-5}$
Find: Domain, range, and inverse of $f(x)$
And: Domain and range of $f^{-1}(x)$
Then: Sketch both $f(x)$ & $f^{-1}(x)$ on same graph.

$y = f(x)$	D: $(-\infty, \infty)$	R: $(-\infty, \infty)$
Find the Inverse of $f(x)$	$y = \sqrt[3]{x-5}$ $x = \sqrt[3]{y-5}$ $x^3 = y - 5$ $\rightarrow$ $f^{-1}(x) = x^3 + 5$	Original Equation Switch x and y
Let $g = f^{-1}(x)$ $g = x^3 + 5$	D: $(-\infty, \infty)$ R: $(-\infty, \infty)$ Domain & Inverses are switched. They're the same, so not a big deal.	
Graph $y = f(x)$ $g = f^{-1}(x)$	$g = x^3 + 5$ $y = \sqrt[3]{x-5}$	

Logarithms

Logarithm Properties	
Definition of a Log	$log_b(a) = n \iff b^n = a$
Reference Points $(1,0)$ & $(b,1)$	$log_b(a) = n \iff b^n = a$
	$log_b(b) = 1 \iff b^1 = b$
Conventions	$log_{10}(a) = \log(a)$ $log_e(a) = \ln(a)$
Product Rule	$\log(a) + \log(b) = \log(ab)$
Quotient Rule	$\log(a) - \log(b) = \log\left(\frac{a}{b}\right)$
Power Rule	$\log(a^n) = n \cdot \log(a)$
Reciprocal Rule	$\log\left(\frac{1}{a}\right) = \log(a^{-1}) = -\log(a)$
Inverse Properties	$log_b(b^x) = x$
	$b^{log_b(x)} = x$
Change of Base	$log_b(a) = \frac{log_c(a)}{log_c(b)} = \frac{\log(a)}{\log(b)}$

Logarithm Graphs

$y = \log_b(x)$ $b > 1$	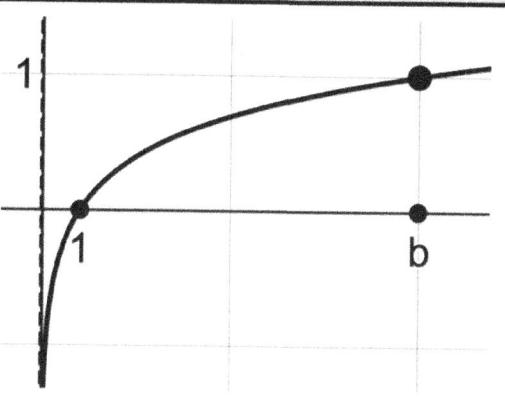
$y = \log_b(x)$ $b < 1$	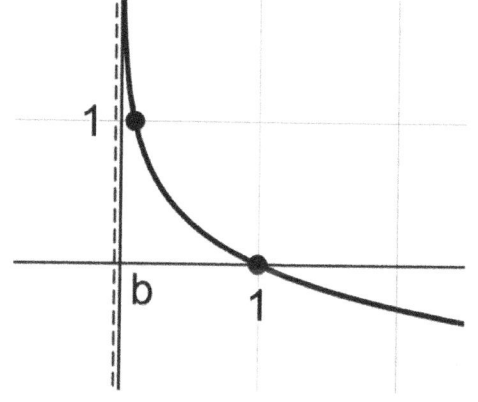

Domain: $(0, \infty)$

Range: $(-\infty, \infty)$

Asymptote: $x = 0$

$\log_b(1) = 0 \qquad \log_b(b) = 1$

Logarithms -- Ex. 1

Use the definition of a log to convert the following logs to exponential form.

$$\log_b(a) = n \quad \Leftrightarrow \quad b^n = a$$

Log	Exponential Form
$\log_9(81) = 2$	$9^2 = 81$
$\ln 1 = 0$	$e^0 = 1$
$\log 1 = 0$	$10^0 = 1$
$\log_7(1) = 0$	$7^0 = 1$
$\log_b(12) = 34$	$b^{34} = 12$
$\log_2(8) = 3$	$2^3 = 8$
$\log_q(r) = s$	$q^s = r$
$\log 10 = 1$	$10^1 = 10$
$\ln e = 1$	$e^1 = e$
$\log_b(b) = 1$	$b^1 = b$

Logarithms -- Ex. 2
Use the definition of a log to convert the following from exponential form to log form.

Exponential Form	Log Form
$5^3 = 125$	$\log_5(125) = 3$
$\left(\frac{1}{5}\right)^{-3} = 125$	$\log_{\frac{1}{5}}(125) = -3$
$10^5 = 100{,}000$	$\log(100{,}000) = 5$
$b^0 = 1$	$\log_b(1) = 0$
$b^9 = n$	$\log_b(n) = 9$
$2^3 = 8$	$\log_2(8) = 3$
$10^1 = 10$	$\log 10 = 1$
$e^1 = e$	$\ln e = 1$
$b^1 = b$	$\log_b(b) = 1$
$12^n = 34$	$\log_{12}(34) = 12$
$10^n = 123$	$\log 123 = n$

Logarithms -- Ex. 3

Simplify the expressions without using a calculator.

Recall: $\log(a^n) = n \cdot \log(a)$ and $\log_b(b) = 1$

Expression	Simplified
$\log_3(9)$	$\log_3(3^2)$ $2 \cdot \log_3(3)$ $2 \cdot (1) = 2$
$\log_2(16)$	$\log_2(2^4)$ $4 \cdot \log_2(2)$ $4 \cdot (1) = 4$
$\log_2\left(\frac{1}{16}\right)$	$\log_2(2^{-4})$ $-4 \cdot \log_2(2)$ $-4 \cdot (1) = -4$
$\log_{\frac{1}{6}}\left(\frac{1}{36}\right)$	$\log_{\frac{1}{6}}\left(\left(\frac{1}{6}\right)^2\right)$ $2 \cdot \log_{\frac{1}{6}}\left(\frac{1}{6}\right)$ $2 \cdot (1) = 2$

Graphs of Log Functions -- Ex. 4a

For each given function: Graph it and identify the VA. Also state the domain and range in interval notation.

<u>Hint:</u> Plot points $(1, 0)$ & $(b, 1)$ to help with graphing.

$y = \log_3(x + 2)$

H. Shift left by 2
No V. Translations

$x - 2$	x	y	y
-1	1	0	0
1	3	1	1

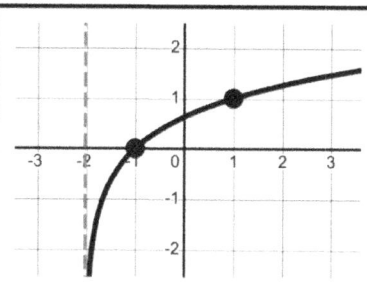

D: $(-2, \infty)$
R: $(-\infty, \infty)$
VA: $x = -2$

$y = \log_3(x - 1) - 2$

H. Shift right by 1
V. Shift down by 2

$x + 1$	x	y	$y - 2$
2	1	0	-2
4	3	1	-1

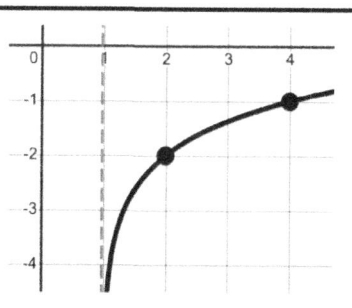

D: $(1, \infty)$
R: $(-\infty, \infty)$
VA: $x = 1$

Graphs of Log Functions -- Ex. 4b

For each given function: Graph it and identify the VA. Also state the domain and range in interval notation.
<u>Hint:</u> Plot points $(1, 0)$ & $(b, 1)$ to help with graphing.

$$y = -\log_4 x$$

V. Rotation over x-axis

No H. Translations

x	x	y	$-y$
1	1	0	0
4	4	1	-1

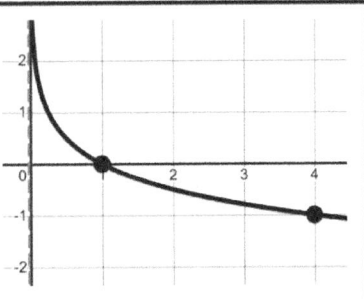

D: $(0, \infty)$
R: $(-\infty, \infty)$
VA: $x = 0$

$$y = \log_2(x - 1)$$

H. Shift right by 1

No V. Translations

$x+1$	x	y	y
2	1	0	0
3	2	1	1

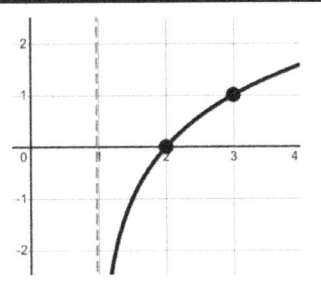

D: $(1, \infty)$
R: $(-\infty, \infty)$
VA: $x = 1$

Domains of Log Functions -- Ex. 5a

For each function write the domain in interval notation.

Function	Calculations	Domain
$y = \log_3(x - 7)$	$x - 7 > 0$ $x > 7$	$(7, \infty)$
$y = \log_2(x^2 + 5)$	$x^2 + 5 > 0$ $x^2 > -5$ True for all x	$(-\infty, \infty)$
$y = \log_7(x^2 - 9)$	$x^2 - 9 > 0$ $x^2 > 9$ $x < -3$ or $x > 3$	$(-\infty, -3)$ U $(3, \infty)$
$y = \log_{\frac{1}{6}}(x + 2)$	$x + 2 > 0$ $x > -2$	$(-2, \infty)$
$y = \ln(x^2)$	$x^2 > 0$ $x \neq 0$	$(-\infty, 0)$ U $(0, \infty)$
$y = \log(x + 2) + 1$	$x + 2 > 0$ $x > -2$	$(-2, \infty)$

Domains of Functions -- Ex. 5b

For each function write the domain in interval notation.

Function	Calculations	Domain
$y = \log(x)$	$x > 0$	$(0, \infty)$
$y = \sqrt{x}$	$x \geq 0$	$[0, \infty)$
$y = \dfrac{1}{x}$	$x \neq 0$	$(-\infty, 0)$ U $(0, \infty)$
$y = \dfrac{1}{\sqrt{x}}$	$x \geq 0$ AND $x \neq 0$ Therefore: $x > 0$	$(0, \infty)$

Log Sums and Differences -- Ex. 6

Write the logarithm as a sum or difference of logs. Simplify as much as possible.

Logarithm	Written as sums & differences.
$\log_a a^2$	$2 \log_a a$ $2(1) = 2$
$\log_4(AB)$	$\log_4 A + \log_4 B$
$\log_4 \left(\frac{1}{16} x^3 z\right)$	$\log_4 \left(\frac{1}{16}\right) + \log_4 x^3 + \log_4 z$ $\log_4(4^{-2}) + \log_4 x^3 + \log_4 z$ $-2 \log_4(4) + 3 \log_4 x + \log_4 z$ $-2 + 3 \log_4 x + \log_4 z$
$\log \left(\dfrac{100}{\sqrt{a^2 + b^2}}\right)$	$\log(100) - \log(\sqrt{a^2 + b^2})$ $\log(10^2) - \log(a^2 + b^2)^{\frac{1}{2}}$ $2 \log(10) - \frac{1}{2} \log(a^2 + b^2)$ $2 - \frac{1}{2} \log(a^2 + b^2)$
$\log \left(\frac{AB}{CD}\right)$	$\log A + \log B - \log C - \log D$

Log Sums and Differences -- Ex. 7
Write the logarithm expression as a single log with a coefficient of 1. Simplify as much as possible.

Logarithm	Written as a single log.
$\ln x + \ln 4$	$\ln(4x)$
$\log_6 144 - \log_6 4$	$\log_6\left(\frac{144}{4}\right)$ $\log_6(36)$ $\log_6(6^2)$ $2 \cdot \log_6(6) = 2$
$5\log_3 x + \log_3 z$	$\log_3(x^5) + \log_3(z)$ $\log_3(x^5 z)$
$\frac{1}{2}\ln(x^2 - 1) - \frac{1}{2}\ln(x + 1)$	$\frac{1}{2}\left[\ln\left(\frac{x^2-1}{x+1}\right)\right]$ $\frac{1}{2}\left[\ln\left(\frac{(x-1)(x+1)}{x+1}\right)\right]$ $\frac{1}{2}[\ln(x-1)]$ $\ln(x-1)^{\frac{1}{2}}$ $\ln\sqrt{x-1}$

Exponential Equations -- Ex. 8

Solve the exponential equations.

Exponential Equation	Solution
$3^x = 81$	$3^x = 81$ $3^x = 3^4$ $x = 4$
$5^{2x+3} = 625$	$5^{2x+3} = 625$ $5^{2x+3} = 5^4$ $2x + 3 = 4$ $x = \frac{1}{2}$
$8^{10x} = 5^{100}$	$8^{10x} = 5^{100}$ $\log(8^{10x}) = \log(5^{100})$ $(10x)\log(8) = (100)\log(5)$ $\frac{(10x)}{(100)} = \frac{\log(5)}{\log(8)} = \log_8 5 \approx .774$ $\frac{x}{10} \approx .774$ $x \approx 7.74$
$4^x = 24$	$\log_4(24) = x$ Definition of a log.

Solving Log Equations -- Ex. 9a

Solve the logarithmic equations.

$log_4(x + 1) = 3$

$4^3 = x + 1$	Definition of a Log

$x + 1 = 64$

$x = 63$

$\log(3x + 11) = \log(3 - x)$

$3x + 11 = 3 - x$

$4x = -7$

$x = -\dfrac{7}{4}$

$log_4(x^2 + 30x) = 3$

$4^3 = x^2 + 30x$	Definition of a Log

$x^2 + 30x - 64 = 0$

$(x - 2)(x + 32) = 0$

$x = 2, -32$

Solving Log Equations -- Ex. 9b

Solve the logarithmic equations.

$5 \log_3(7 - 5x) + 2 = 17$

$5 \log_3(7 - 5x) = 15$

$\log_3(7 - 5x) = 3$

$3^3 = (7 - 5x)$

$5x = 7 - 3^3$

$x = \dfrac{7-27}{5} = \dfrac{-20}{5} = -4$

$\log x + \log(x - 10) = \log(x - 18)$

$\log(x(x - 10)) = \log(x - 18)$

$x^2 - 10x = x - 18$

$x^2 - 11x + 18 = 0$

$(x - 2)(x - 9) = 0$

$x = 2, 9$

Exponential Growth and Decay

Exponential Functions: $y = b^x$

Exponential Growth $y = b^x$ $b > 1$	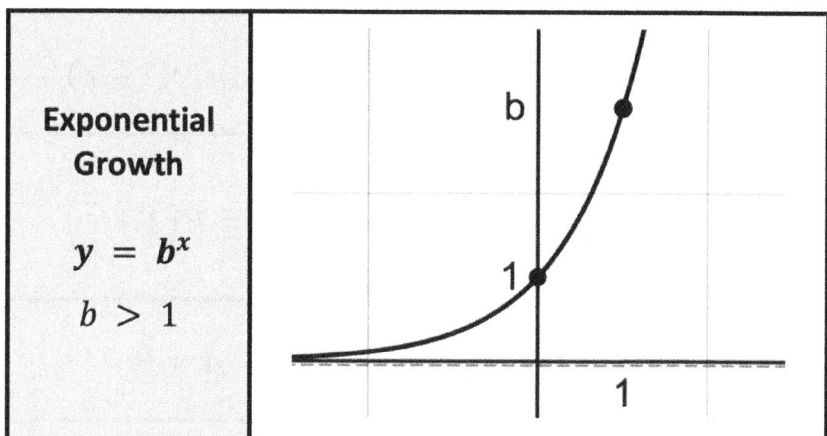

Exponential Decay $y = b^x$ $b < 1$	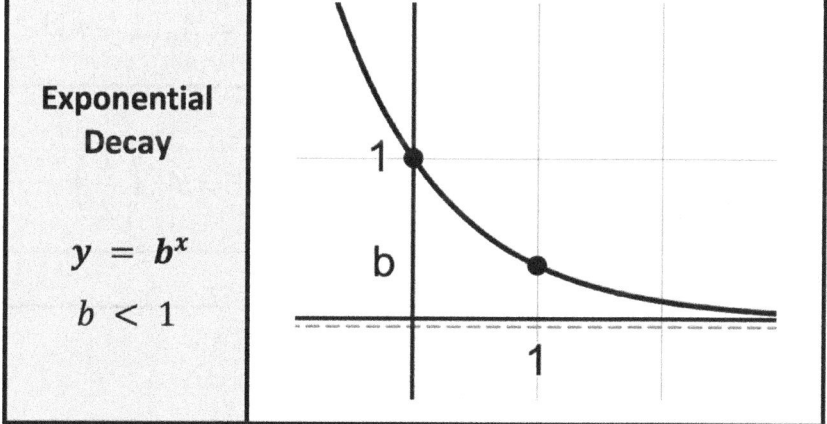

Reference Points:	$(0, 1)$ & $(1, b)$

Exponential Functions: $y = ab^x$

Growth of Money	Compounded Yearly	$A = P(1 + r)^t$
	Compounded n times per year	$A = P\left(1 + \dfrac{r}{n}\right)^{nt}$
	Compounded Continuously	$A = Pe^{rt}$

Decay of a Value	Depreciates at a yearly rate, r	$A = A_0(1 - r)^t$
	Depreciates by half every h years	$A = A_0\left(\dfrac{1}{2}\right)^{\frac{t}{h}}$

Terms	t = Time (in years) r = rate (yearly rate) A = Amount at time t P = Principal (Initial Value) A_0 = Initial Value

Exponential Growth – Ex. 1

If $80,000 is invested in a bank, find the value of the investment under the following conditions.

Conditions	Result
3% interest, Compounded annually, For 50 years.	$A = P(1+r)^t$ $A = 80{,}000(1+0.03)^{50}$ $A = 80{,}000(1.03)^{50}$ $A = \$350{,}712.50$
3% interest, Compounded monthly, For 50 years.	$A = P\left(1+\dfrac{r}{n}\right)^{nt}$ $A = 80{,}000\left(1+\dfrac{.03}{12}\right)^{12(50)}$ $A = 80{,}000(1.025)^{600}$ $A = \$350{,}864.60$
3% interest, compounded continuously, For 50 years.	$A = Pe^{rt}$ $A = (80{,}000)\,e^{(.03)(50)}$ $A = (80{,}000)\,e^{1.5}$ $A = \$358{,}535.10$

	Exponential Decay – Ex. 2
Joe bought a new car for $60,000. If the value of the car depreciates at a rate of 6% per year, how much will his car be worth in ten years? In 15 years? When will his car be worth $10,000 ?	
Value of car in 10 years	$A = A_0(1-r)^t$ $A = 60{,}000\,(1-.06)^{10}$ $A = 60{,}000\,(0.94)^{10}$ $A = \$32{,}316.90$
Value of car in 15 years	$A = A_0(1-r)^t$ $A = 60{,}000\,(1-.06)^{15}$ $A = 60{,}000\,(0.94)^{15}$ $A = \$23{,}717.50$
When car is worth $10,000	$A = A_0(1-r)^t$ $10{,}000 = 60{,}000\,(1-.06)^t$ $0.16667 = (0.94)^t$ $\log(0.16667) = \log(0.94)^t$ $\log(0.16667) = t \cdot \log(0.94)$ $t = \dfrac{\log(0.16667)}{\log(0.94)} \approx 29\ yrs.$

Exponential Growth (Unknown Rate) – Ex. 3

Sally, the scientist, is growing a new kind of algae in a laboratory. She started with 1000 cells. In 1 hour, there were 1234 algae cells. How many algae cells will there be in 5 hours, after she started?
How long will it take the number of algae to triple?

Determine the growth rate.	$r = \dfrac{Change}{Original} = \dfrac{new - old}{old}$ $r = \dfrac{1234 - 1000}{1000} = \dfrac{234}{1000} = .234$ $r = 23.4\%$ increase per hour
Quantity in 5 hours	$A = A_0(1+r)^t$ $A = 1000 \, (1 + .234)^5$ $A = 1000 \, (1.234)^5$ $A = 2861.38 \approx 2861$ algae
When number of algae triples	$A = A_0(1+r)^t$ $3000 = 1000 \, (1 + .234)^t$ $3 = (1.234)^t$ $\log(3) = \log 1.234^t$ $\log(3) = t \cdot \log(1.234)$ $t = \dfrac{\log(3)}{\log(1.234)} \approx 5.2$ hours

Half Life – Ex. 4

Pd-100 has a half-life of 3.6 days. Sam, the scientist, started with one mole of this substance.

Note: $1\ mole = 6.023 \times 10^{23}\ atoms$

(1) How many atoms will be present after 15 days?
(2) When will the number of atoms be 6 million?

Quantity after 15 days	$A = A_0 \left(\dfrac{1}{2}\right)^{\frac{t}{h}}$ $A = 6.023 \times 10^{23} \left(\dfrac{1}{2}\right)^{\frac{15}{3.6}}$ $A = 3.35\ E\ 22 = 3.35 \times 10^{22}$ atoms
When 6 million atoms remain.	$A = A_0 \left(\dfrac{1}{2}\right)^{\frac{t}{h}}$ $6{,}000{,}000 = 6.023 \times 10^{23} \left(\dfrac{1}{2}\right)^{\frac{t}{3.6}}$ $9.962 \times 10^{-18} = (.5)^{\frac{t}{3.6}}$ $\log(9.962 \times 10^{-18}) = \left(\dfrac{t}{3.6}\right) \log(.5)$ $\dfrac{t}{3.6} = \dfrac{\log(9.962 \times 10^{-18})}{\log(0.5)} = 56.48$ $t = (56.48)(3.6) \approx 203$ days

Regression

Regression

Regression, in math, is fitting a curve, $y = f(x)$, to a set of points. Mathematical techniques can be used, but are not discussed in this section.

Here, regression will be demonstrated in two ways:
1) With the Desmos, online graphing tool, and
2) With a TI graphing calculator.

After that, regression will be used to find the time until the next eruption of the "Old Faithful" geyser at Yellowstone National Park.

Type of Regression should match the data. Graph the data to see what type of regression will be best.
- Linear Regression   (straight line)
- Quadratic Regression   (parabolic shape)
- Exponential Regression (exponential shape)
- Etc.

Goodness of Fit:
- R and R^2 values tell how well the regression matches the set of points.
- The closer $|R|$ and R^2 are to 1, the better.
- If $R^2 = 1$ it's a perfect fit.

Regression on Desmos (Desmos.com)

Regression on Desmos can be done on:
- Using the Desmos app, on your mobile phone.
- Using the Desmos website, on you laptop.

The following set of points will be used to demonstrate regression on the Desmos website.

x	0	1	2	3	4	5	6
y	6	2	−.5	−2.5	0	2	8

- Click the + sign, top left.
- Select "Table"
- Input all data points

x_1	y_1
0	6
1	2
2	−.5
3	−2.5

Adjust graph settings to fit your data.
Click on wrench.

Plot the points.
Type this: (x1, y1)
Desmos formats subscripts.

(x_1, y_1)

Regression on Desmos (Continued)	
Looks like quadratic regression is appropriate.	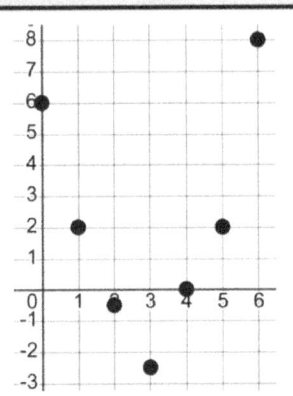
Type this: $y_1 \sim ax_1^2 + bx_1 + c$ Notes: • Just type y1 to get y_1 • Just type x1 to get x_1 • Use " ~ " (not " = ")	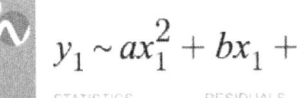 $y_1 \sim ax_1^2 + bx_1 + c$ STATISTICS RESIDUALS $R^2 = 0.9785$ e_1 plot PARAMETERS $a = 0.970238$ $b = -5.58929$ $c = 6.29762$
Graph shows regression curve and data points.	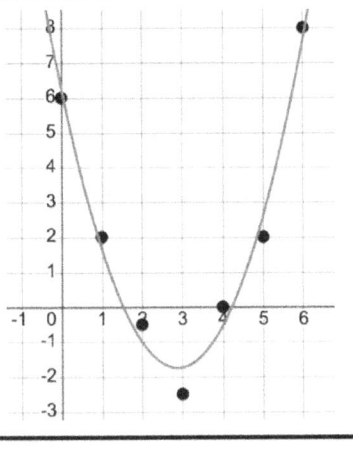

Regression on Desmos -- Using the Equation

Use the quadratic regression to find " y " when $x = 12$. Do this in two ways:
1) Using the equation
2) Using the graph

Previously, Desmos was used to fit a quadratic function $y = .97x^2 - 5.59x + 6.30$ to these points:

x	0	1	2	3	4	5	6
y	6	2	-.5	-2.5	0	2	8

Find " y " when $x = 12$ Use equation.	$y = .97x^2 - 5.59x + 6.30$ $y = .97(12)^2 - 5.59(12) + 6.30$ $y = 78.90$
Find " y " when $x = 12$ Use graph. Graph on next page.	• Enter equation: $y = .97x^2 - 5.59x + 6.30$ • Enter equation: $x = 12$ • Expand window to show intersection. • Click on intersection to shop point $(12, 78.9)$.

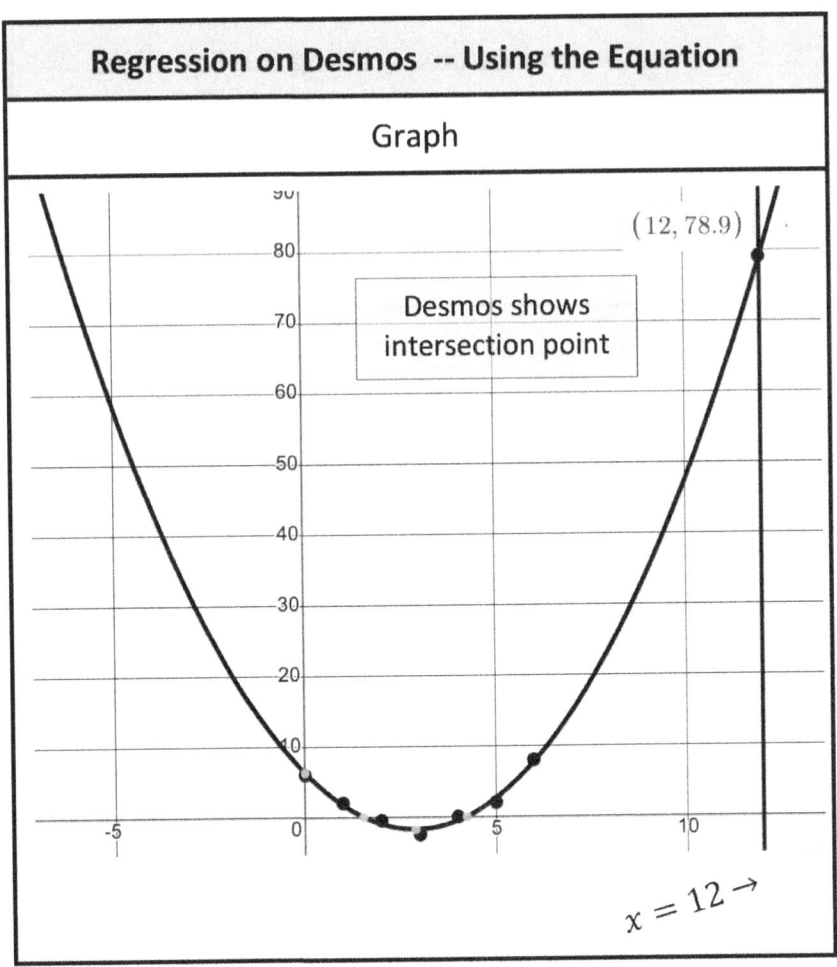

Regression on TI Graphing Calculator

The following set of points will be used to demonstrate regression on the Desmos website.

x	0	1	2	3	4	5	6
y	6	2	−.5	−2.5	0	2	8

Input data points into a table.	• Press **[stat]**, EDIT, **[Enter]** • Clear data from L1 & L2 • Move cursor to top of column and press **[clear]**. • Input data under L1 & L2.
Adjust graphing window to fit data.	• Press **[window]** • Set x & y. (max & min)
Plot data points. Look at shape to determine type of regression	• Press **[y=]** button • Use up arrow to select **Plot1** to plot points. • Press **[graph]** button.
Points are plotted. Looks like Quadratic regression Is appropriate.	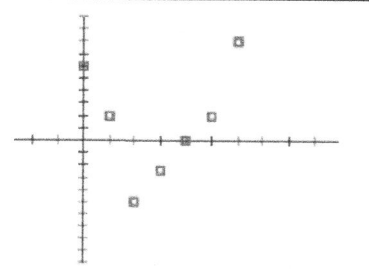

Regression on TI Graphing Calculator	
Continued …	
Turn on diagnostics to get R & R^2 Values	• Press **[2nd]** , **[0]** to select **Catalog** . • Scroll down and select "DiagnosticsOn" • ENTER, ENTER, Done.
Do regression on your data.	• Press **[stat]**, CALC. • Scroll down to select "QuadReg", **[enter]**
Select regression information (No Graph)	• Take all defaults. • Scroll down to Calculate.
Select regression information (With Y1 Graph)	• Scroll to "Store RegEQ:" • **[vars]**, Y-VARS , Function • **[enter]** , Y1, **[enter]** • Scroll down to Calculate
Screen should look like this.	QuadReg Xlist:L$_1$ Ylist:L$_2$ FreqList: Store RegEQ:Y$_1$

Regression on TI Graphing Calculator	
Continued ...	
Regression Information is given	QuadReg y=ax²+bx+c a=1.130952381 b=-6.392857143 c=5.976190476 R²=0.9049546944
Graph the Regression equation	• Press [y=], [graph]
The Graph and Points	

To connect your TI-84 graphing calculator to your computer, install the free TI Connect software application via "education.ti.com/ticonnect" on your computer.

Regression on TI Graphing Calc. -- Using the Equation
Use the quadratic regression to find " y " when $x = 12$. Do this in two ways: 1) Using the equation 2) Using the graph
Previously, Desmos was used to fit a quadratic function $y = 1.1x^2 - 6.4x + 6$ to these points: 	x	0	1	2	3	4	5	6
---	---	---	---	---	---	---	---
y	6	2	−.5	−2.5	0	2	8	

Find " y " when $x = 12$ Use equation.	$y = 1.13x^2 - 6.39x + 5.98$ $y = 1.13(12)^2 - 6.39(12) + 5.98$ $y = 92.02$
Find " y " when $x = 1$ Use graph. Graph on next page.	• Press **[window]**, expand window • Press **[graph]** • Press **[trace]** • Up arrow toggles between the plot (points) and Y1 (function) Selection displayed along top • Press up arrow to select Y1 • Type: 12

Regression on TI Graphing Calc. -- Using the Equation
Graph
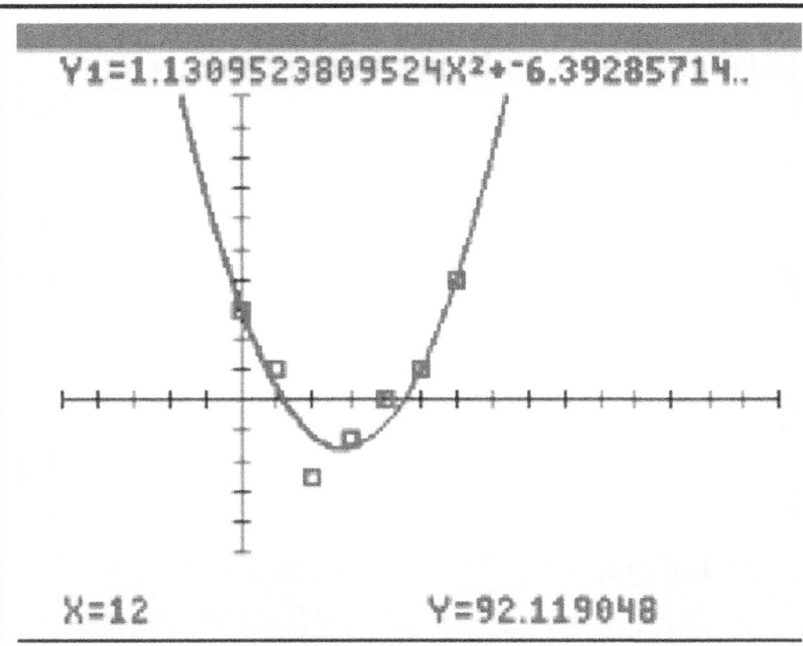
When in **[trace]** mode, just type the value of X And the value of Y will be displayed.

To connect your TI-84 graphing calculator to your computer, install the free TI Connect software application via "education.ti.com/ticonnect" on your computer.

Old Faithful Geyser – Problem Statement

True story: My family's tour at Yellowstone National Park, included a stop at the Old Faithful geyser. The Old Faithful geyser is famous because it erupts on a relatively predictable schedule. On average, the geyser erupts for about 3.5 minutes. Then, the next eruption occurs about 71 minutes later. In general, if the geyser eruption is short, then the time until the next eruption is also shorter and vice-versa.

Our tour-guide made sure we stopped at Old Faithful before the next predicted eruption.

I decided to look online to find some data about the Old Faithful geyser and use regression to develop my own predictor.

Problem Statement: Find and use online data for the Old Faithful geyser to predict the time until the next eruption.

Old Faithful Geyser Regression – Getting Data

Find and use online data for the Old Faithful geyser to predict the time until the next eruption.

There are many sources of online data. I just picked one that looked reasonable and reliable. I found some data from Carnegie Mellon University (CMU) that included 273 records of:
- Eruption Time
- Time to Next Eruption

Data was copied from online and pasted into an Excel Spreadsheet. The first few records are shown.

	A
1	1 , 3.600 , 79
2	2 , 1.800 , 54
3	3 , 3.333 , 74
4	4 , 2.283 , 62

Within Excel, data was separated and formatted. If necessary, search online for how to do this.

	A	B	C
1	1	3.6	79
2	2	1.8	54
3	3	3.333	74
4	4	2.283	62
5	5	4.533	85

Old Faithful Geyser Regression – Formatting Data

Find and use online data for the Old Faithful geyser to predict the time until the next eruption.

Make sure you have all the data you need.
What is your output? What are your input(s)?

Output	D. Time until next eruption.
Input(s)	The input could be the time of the last eruption. Or, perhaps the prediction would be better if more inputs were used. These inputs were considered: A. Previous duration of eruption B. Previous time until next eruption. C. Duration of latest eruption

For these Values, Averages were used.

	A	B	C	D
1	Dur Prev	Next Prev	Duration	Next
2	3.475	71.000	3.6	79
3	3.600	79.000	1.8	54
4	1.800	54.000	3.333	74
5	3.333	74.000	2.283	62
6	2.283	62.000	4.533	85

Old Faithful Geyser Regression – Load Desmos Table

Find and use online data for the Old Faithful geyser to predict the time until the next eruption.

Go to Desmos.com and click on the "Graphing" tools. Click on the "+" then "f(x) expression."

Note: "Table" is for 2-columns. "f(x) expression" is for multiple columns.

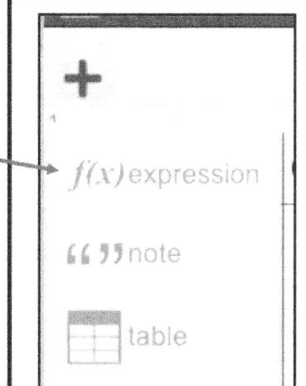

Copy first 4 columns in Excel. Paste into Desmos.

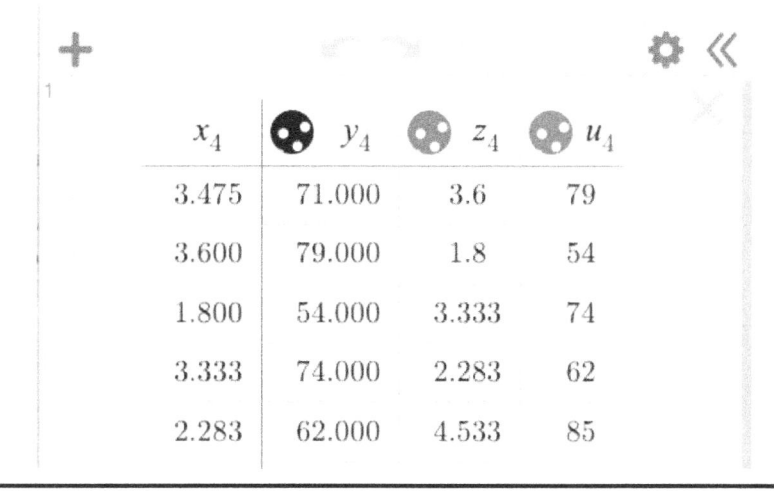

x_4	y_4	z_4	u_4
3.475	71.000	3.6	79
3.600	79.000	1.8	54
1.800	54.000	3.333	74
3.333	74.000	2.283	62
2.283	62.000	4.533	85

Old Faithful Geyser Regression – Do Regression

Prev Dur x_4	Prev Next y_4	Dur z_4	Next u_4
3.475	71.000	3.6	79
3.600	79.000	1.8	54
1.800	54.000	3.333	74

Two regressions were done to predict "Next" eruption

$u_4 \sim az_4 + b$

STATISTICS RESIDUALS

$r^2 = 0.8115$ e_1
$r = 0.9008$

PARAMETERS

$a = 10.7296$ $b = 33.4744$

$u_4 \sim ax_4 + by_4 + cz_4 + d$

STATISTICS RESIDUALS

$R^2 = 0.8263$ e_4 plot

PARAMETERS

$a = 3.11438$ $b = -0.282415$ $c = 10.5745$
$d = 43.1838$

Old Faithful Geyser Regression – Conclusion

Find and use online data for the Old Faithful geyser to predict the time until the next eruption.

Compare regression with 1 input to the regression with 3 inputs.	The regression with 1 input is more practical and had $r^2 = 0.81$ The regression with 3 inputs is more complex and had $R^2 = 0.83$ (not much better)
Conclusion	The regression with 1 input is more practical and simple to use. The additional accuracy of the 3-input regression is not worth the trouble!
Regression based on one input: Duration of most recent eruption.	$y = 10.7x + 33.5$ $y =$ Time to next eruption. $x =$ Duration of last eruption.

Systems of Linear Equations

Systems of Linear Equations in 2 Variables

A **System of Linear Equations** is a set of linear equations in the same two variables. When two linear systems are graphed, there are three possibilities, as shown below.

Possibility	Example	Graph
Intersecting Lines 1 Solution	$y = x - 3$ $y = -x + 5$	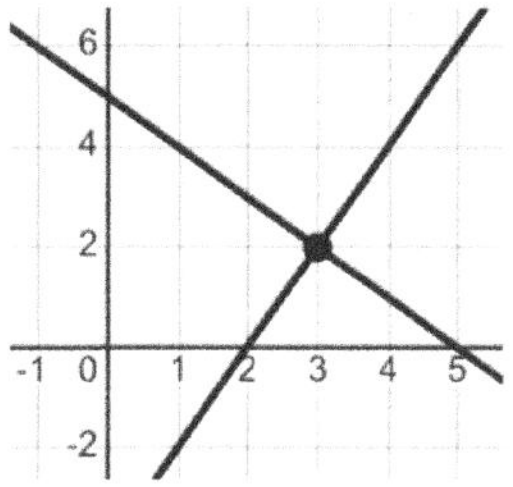
Parallel Lines 0 Solutions	$y = \frac{x}{2} + 3$ $y = \frac{x}{2} - 2$	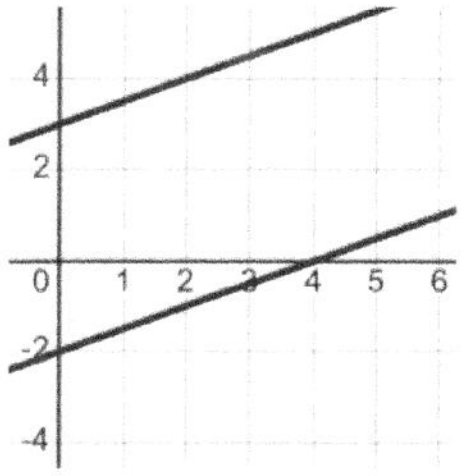
Same Lines ∞ Solutions	$y - 2 = x$ $y = x + 2$	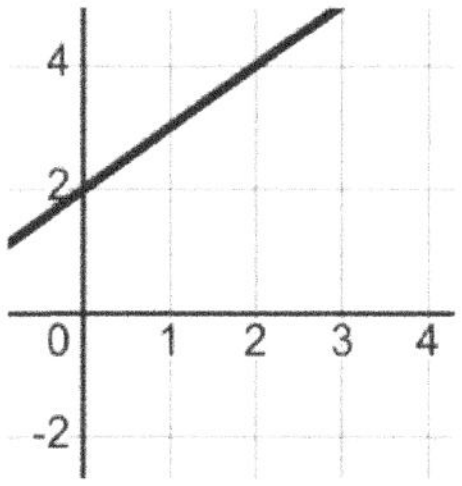

Systems of Linear Equations in 2 Variables
How to Solve

Algebraic Solutions

Substitution	Elimination
• Isolate one variable • Substitute it into the other equation. • That equation will have only one variable. • Solve it to get one variable. • Use known variable to find the other.	• Multiply equation(s) so one variable has the same coefficient in both equations. • Add both equations to eliminate one variable. • Solve it to get one variable. • Use known variable to find the other.

Graphical Solution

- Graph both equations on the same coordinate axis.
- Find the point where they intersect.
- The coordinates of the intersection is the solution.

Systems of Linear Equations – Ex. 1a

Given this system: $\begin{cases} x - 2y = 5 \\ 4x + 3y = 9 \end{cases}$

Solve using substitution, elimination, and graphing.

Substitution Solution	
Isolate one variable	$x - 2y = 5$ $x = 5 + 2y$
Substitute into other equation and solve.	$4x + 3y = 9$ $4(5 + 2y) + 3y = 9$ $20 + 8y + 3y = 9$ $11y = -11$ $y = -1$
Use known variable to find the other.	$x - 2y = 5$ $x - 2(-1) = 5$ $x + 2 = 5$ $x = 3$
Solution	$(x, y) = (3, -1)$

Systems of Linear Equations – Ex. 1b

Given this system: $\begin{cases} x - 2y = 5 \\ 4x + 3y = 9 \end{cases}$

Solve using substitution, elimination, and graphing.

	Elimination Solution	
Multiply eqns. So one variable has matching coefficients.	$x - 2y = 5$ $4x + 3y = 9$	$\rightarrow$ $4x - 8y = 20$ $4x + 3y = 9$
Subtract 2nd from 1st eqn. and solve for one variable.	$4x - 8y = 20$ $-(4x + 3y = 9\ )$ $-11y = 11$ $y = -1$	
Use known variable to find the other.	$x - 2y = 5$ $x - 2(-1) = 5$ $x + 2 = 5$ $x = 3$	
Solution	$(x, y) = (3, -1)$	

Systems of Linear Equations – Ex. 1c

Given this system: $\begin{cases} x - 2y = 5 \\ 4x + 3y = 9 \end{cases}$

Solve using substitution, elimination, and graphing.

Graphing Solution

$x - 2y = 5$	$4x + 3y = 9$
$x = 0 \rightarrow y = -\frac{5}{2}$	$x = 0 \rightarrow y = 3$
$y = 0 \rightarrow x = 5$	$y = 0 \rightarrow x = \frac{9}{4}$

Intersection: $(x, y) = (3, -1)$

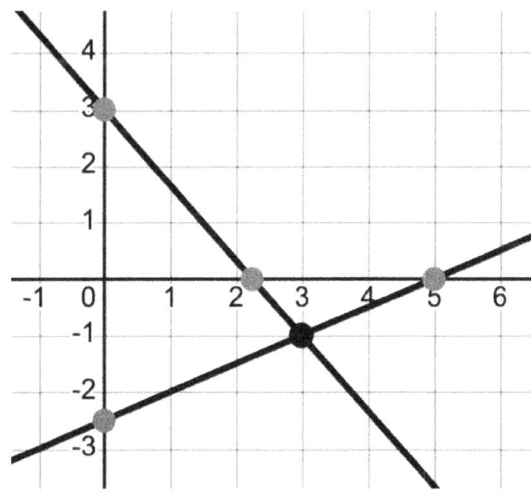

Systems of Linear Inequalities

Systems of Linear Inequalities in 2 Variables

A **System of Linear Inequalities** is a set of linear inequalities in the same two variables.

To solve a system of linear inequalities, do this:
- Graph both on the same coordinate axis.
- Shade one side of each to show the feasible region.
- The solution is the set of ordered pairs, within the intersection of the shaded regions.

Examples

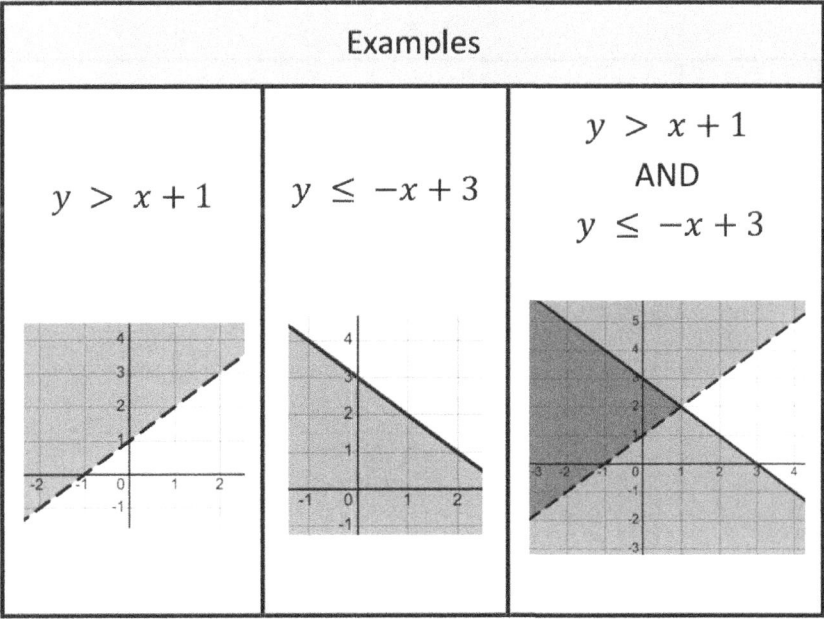

| $y > x + 1$ | $y \leq -x + 3$ | $y > x + 1$ AND $y \leq -x + 3$ |

Systems of Linear Inequalities in 2 Variables
Ex. 1a

Given this system of inequalities:

$$y < -\frac{1}{2}x + 4$$
$$y \geq x - 2$$

1) Graph the inequalities to show the feasible region.

2) Determine if the following points are within the feasible region: $(1,1), (4,1), (6,1), (6,4)$

Graph

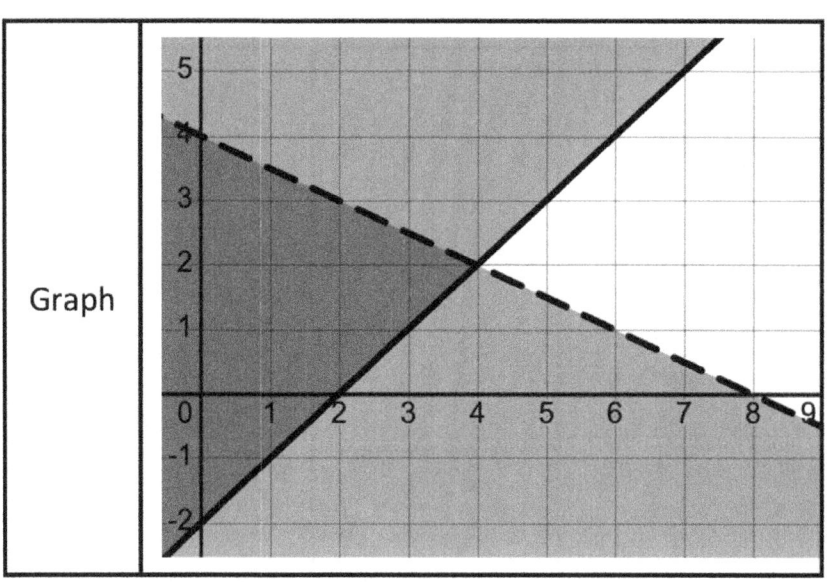

Systems of Linear Inequalities in 2 Variables
Ex. 1b

Given this system of inequalities:
$$y < -\frac{1}{2}x + 4$$
$$y \geq x - 2$$

1) Graph the inequalities to show the feasible region.

2) Determine if the following points are in the feasible region: $(1,1), (2,3), (3,1), (4,1), (6,1), (6,4)$

(1, 1)　Yes

(2,3)　No

(3, 1)　Yes

(4,1)　No

(6,1)　No

(6,4)　No

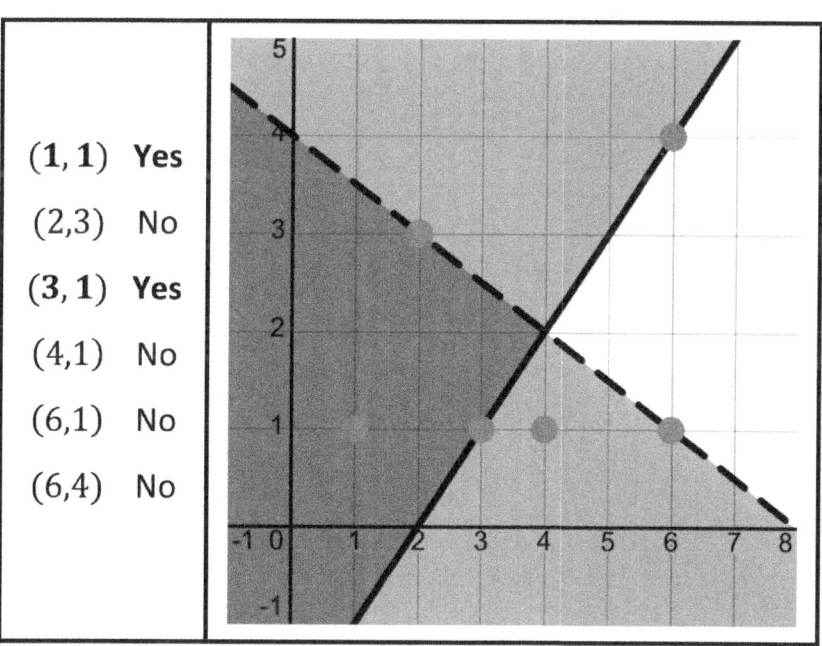

Solving Inequalities

	Solving Inequalities **With the Critical Value Method (CVM)**
	The **Critical Value Method (CVM)** is a process used to solve any inequality – simple or complex. The process is outlined below.
1.	Identify restrictions on x, if any.
2.	Find critical values. • Change the inequality sign to an equal sign. The solutions are critical values. • Restrictions on x (if any) are also critical values.
3.	Sketch a number line, with the critical values. Note: n critical values will create $(n + 1)$ intervals on the number line.
4.	Test one value, in each interval, using the original inequality. Do not use boundary numbers. • If it makes the original inequality true, then all numbers in that interval are solutions. • If it makes the original inequality false, then all numbers in that interval are NOT solutions.
5.	Solution: Write solution in interval notation.

Critical Value Method (CVM) -- Ex. 1

Given: $|x + 2| \leq 5$ Solve for x

1.	Identify restrictions on x	None.
2.	Find Critical Values Change the inequality sign to an equal sign and solve.	$|x + 2| = 5$ $x + 2 = 5$ $x = 3$ $x + 2 = -5$ $x = -7$ $x = -7, 3$ (Critical Values)
3.	Sketch a number line, with the critical values.	●————● -7 3
4.	Test one value, in each interval, using the original inequality. $|x + 2| \leq 5$	F T F ●————● -7 3 $x = -8 \rightarrow$ False $x = 0 \rightarrow$ True $x = 5 \rightarrow$ False
5.	Solution:	$x = [-7, 3]$

Critical Value Method (CVM) -- Ex. 2

Given: $|x + 2| \geq 5$ Solve for x

1.	Identify restrictions on x	None.
2.	Find Critical Values. Change the inequality sign to an equal sign and solve.	$|x + 2| = 5$ $x + 2 = 5$ $x = 3$ $x + 2 = -5$ $x = -7$ $x = -7, 3$ (Critical Values)
3.	Sketch a number line, with the critical values.	•―――――• -7 3
4.	Test one value, in each interval, using the original inequality. $\|x + 2\| \geq 5$	T F T ――•―――•―― -7 3 $x = -8 \rightarrow$ True $x = 0 \rightarrow$ False $x = 5 \rightarrow$ True
5.	Solution:	$x = (-\infty, -7] \cup [3, \infty)$

	Critical Value Method (CVM) -- Ex. 3	
	Given: $\sqrt{x+2} < 5$	Solve for x.
1.	Identify restrictions on x	$x + 2 \geq 0$ $x \geq -2$
2.	Find Critical Values. Change inequality sign to an equal sign and solve.	$\sqrt{x+2} = 5$ $x + 2 = 5^2$ $x = 23$ (Critical Value)
3.	Sketch a number line, with the critical values.	(number line with open circle at 23)
4.	Test one value, in each interval, using the original inequality. $\sqrt{x+2} < 5$	(number line: T at left of 23, F at right) $x = 22 \rightarrow$ True $x = 24 \rightarrow$ False
5.	Solution: Must consider restrictions on x.	$x = [-2, 23)$

	Critical Value Method (CVM) -- Ex. 4	
	Given: $x^2 + 7x > 0$	Solve for x.
1.	Identify restrictions on x	None.
2.	Find Critical Values. Change inequality sign to an equal sign and solve.	$x^2 + 7x = 0$ $x(x+7) = 0$ $x = 0, -7$
3.	Sketch a number line, with the critical values.	○———○ -7 0
4.	Test one value, in each interval, using the original inequality. $x^2 + 7x > 0$	T F T —○——○— -7 0 $x = -8 \rightarrow$ True $x = -4 \rightarrow$ False $x = 2 \rightarrow$ True
5.	Solution:	$x = (-\infty, -7) \cup (0, \infty)$

Critical Value Method (CVM) -- Ex. 5	
Given: $\dfrac{x-4}{x-1} \leq 0$ Solve for x.	
1. Identify restrictions on x	$x \neq 1$ (Critical Value)
2. Find Critical Values. Change inequality sign to an equal sign and solve.	$\dfrac{x-4}{x-1} = 0$ $x - 4 = 0$ $x = 4$ (Critical Value)
3. Sketch a number line, with the critical values.	number line with open circle at 1 and closed circle at 4
4. Test one value, in each interval, using the original inequality. $\dfrac{x-4}{x-1} \leq 0$	number line with F, T, F regions $x = 0 \rightarrow$ False $x = 2 \rightarrow$ True $x = 5 \rightarrow$ False
5. Solution:	$x = (1, 4]$

	Critical Value Method (CVM) -- Ex. 6	
Given: $\frac{3x+1}{x-2} \geq 4$		Solve for x.
1.	Identify restrictions on x	$x \neq 2$ (Critical Value)
2.	Find Critical Values. Change inequality sign to an equal sign and solve.	$\frac{3x+1}{x-2} = 4$ $3x + 1 = 4x - 8$ $x = 9$ (Critical Value)
3.	Sketch a number line, with the critical values.	──○────●──── 2 9
4.	Test one value, in each interval, using the original inequality. $\frac{3x+1}{x-2} \geq 4$	F T F ──○────●──── 2 9 $x = 0 \rightarrow$ False $x = 3 \rightarrow$ True $x = 10 \rightarrow$ False
5.	Solution:	$x = (2, 9]$

Algebra Topics – Part 2

Roots

Roots = Zeros

Graphing polynomial functions is easy if the polynomial is in factored form. And the zeros of a polynomial function are easy to find if the polynomial is in factored form:

$$f(x) = a(x-z_1)^i (x-z_2)^j \ldots (x-z_n)^k$$

Here, the zeros are: $z_1, z_2, \ldots, z_n$

If $x = z_1, z_2, \ldots, z_n$ then the function is zero.

If c is a zero of a polynomial, $f(x)$, Then ...	Then $(x-c)$ is a linear factor.
	$f(c) = 0$

Types of Roots	Form	Notes
Rational	$\dfrac{a}{b}$	Find possible Rational Roots with the Rational Root Theorem.
Irrational	$a \pm b\sqrt{c}$	Come in conjugate pairs
Imaginary	$a \pm bi$	Come in conjugate pairs.

Rational Roots

The **Rational Root Theorem** gives us a list of possible rational roots for a polynomial. Recall, a rational number is a number that can be written as a fraction of two integers. The **Rational Root Theorem** says the possible zeros of a polynomial function are: $\pm \frac{p}{q}$ where: q is a factor of the leading coefficient and p is a factor of the last constant term

Example: $f(x) = 2x^2 - 5x - 3$

Factors of last term: $\pm 1, \pm 3$
Factors of leading coefficient: $\pm 1, \pm 2$
Possible Zeros are: $\pm \frac{1}{1}, \pm \frac{3}{1}, \pm \frac{1}{2}, \pm \frac{3}{2}$

Synthetic Division is often used to check for zeros.

$f(x) = 2x^2 - 5x - 3$
$f(x) = (2x + 1)(x - 3)$
$f(x) = 2\left(x + \frac{1}{2}\right)(x - 3)$
Zeros at $x = -\frac{1}{2}, 3$
Both zeros are in the above list of "Possible Zeros."

$f\left(-\frac{1}{2}\right) = 0$ and $f(3) = 0$

Irrational Roots
Irrational Roots are in the form: $a \pm b\sqrt{c}$ If $(a + b\sqrt{c})$ is a zero, then $(a - b\sqrt{c})$ is also a zero. The pair of irrational roots is called a "**Conjugate Pair**."

	Given Zeros: $(2\sqrt{3})$ and $(-2\sqrt{3})$
Linear Factors If c is a zero, $(x - c)$ is a linear factor.	$(x - (2\sqrt{3})) = (x - 2\sqrt{3})$
	$(x - (-2\sqrt{3})) = (x + 2\sqrt{3})$
Function with given zeros.	$f(x) = (x - 2\sqrt{3})(x + 2\sqrt{3})$
Recall The difference of two squares	$a^2 - b^2 = (a - b)(a + b)$
Expand $f(x)$	$f(x) = x^2 - (2\sqrt{3})^2$ $f(x) = x^2 - (2^2)(3)$ $f(x) = x^2 - 12$
Conclusion	$f(x) = x^2 - 12$ Has the given zeros.

Irrational Roots (Continued)

Irrational Roots are in the form: $a \pm b\sqrt{c}$
If $(a + b\sqrt{c})$ is a zero, then $(a - b\sqrt{c})$ is also a zero.
The pair of irrational roots is called a "**Conjugate Pair**."

	Given Zeros: $(1 + 2\sqrt{3})$ and $(1 - 2\sqrt{3})$
Linear Factors	$(x - (1 + 2\sqrt{3})) = (x - 1 - 2\sqrt{3})$
	$(x - (1 - 2\sqrt{3})) = (x - 1 + 2\sqrt{3})$
Function	$f(x) = (x - 1 - 2\sqrt{3})(x - 1 + 2\sqrt{3})$
Expand $f(x)$	$f(x) = x^2 - 2x - 11$ (Do it yourself!)
Check with Quadratic Formula	$x = \dfrac{2 \pm \sqrt{2^2 - 4(1)(-11)}}{2(1)} = \dfrac{2 \pm \sqrt{48}}{2}$ $x = \dfrac{2 \pm \sqrt{16(3)}}{2} = \dfrac{2 \pm 4\sqrt{3}}{2} = 1 \pm 2\sqrt{3}$
Conclusion	$f(x) = x^2 - 2x - 11$ Has the given zeros.

Imaginary Roots
Imaginary Roots are in the form: $a \pm bi$ If $(a + bi)$ is a zero, then $(a - bi)$ is also a zero. These imaginary roots are called a "**Conjugate Pair**."

Given Zeros: $(2i)$ and $(-2i)$	
Linear Factors If c is a zero, $(x - c)$ is a linear factor.	$(x - (2i)) = (x - 2i)$
	$(x - (-2i)) = (x + 2i)$
Function with given zeros.	$f(x) = (x - 2i)(x + 2i)$
Recall The difference of two squares	$a^2 - b^2 = (a - b)(a + b)$
Expand $f(x)$	$f(x) = x^2 - (2i)^2$ $f(x) = x^2 - (2^2)(i^2)$ $f(x) = x^2 - (4)(-1)$ $f(x) = x^2 + 4$
Conclusion	$f(x) = x^2 + 4$ Has the given zeros.

Imaginary Roots (Continued)
Imaginary Roots are in the form: $a \pm bi$ If $(a + bi)$ is a zero, then $(a - bi)$ is also a zero.   These imaginary roots are called a "**Conjugate Pair**."

Given Zeros: $(1 + 2i)$ and $(1 - 2i)$	
Linear Factors	$(x - (1 + 2i)) = (x - 1 - 2i)$
	$(x - (1 - 2i)) = (x - 1 + 2i)$
Function	$f(x) = (x - 1 - 2i)(x - 1 + 2i)$
Expand $f(x)$	$f(x) = x^2 - 2x + 5$ (Do it yourself!)
Check with Quadratic Formula	$x = \dfrac{2 \pm \sqrt{2^2 - 4(1)(5)}}{2(1)} = \dfrac{2 \pm \sqrt{-16}}{2}$ $x = \dfrac{2 \pm \sqrt{(16)(-1)}}{2} = \dfrac{2 \pm 4i}{2} = 1 \pm 2i$
Conclusion	$f(x) = x^2 - 2x + 5$ Has the given zeros.

	Graphing Polynomial Functions -- Review
	Given: $f(x) = 3x^2 - 5x - 2$ Find: Degree, Zeros (x intercepts), Multiplicity, Far-End behavior, and y-intercept. Then sketch it.
Factor it	$f(x) = (3x + 1)(x - 2)$ $f(x) = 3\left(x + \frac{1}{3}\right)(x - 2)$
Degree	$D = 1 + 1 = 2$ (even)
Zeros and Multiplicity	$-\frac{1}{3}$, m1 2, m1 Thru Thru
Far-End Behavior	Positive leading coefficient and even degree. ↖ ↗
y-intercept	y-intercept occurs at $x = 0$ $f(0) = -2$ Point: $(0, -2)$
Graph	

Rational Roots -- Ex. 1a	
Given: $f(x) = 3x^2 - 5x - 2$ Find: Degree, Zeros (x intercepts), Multiplicity, Far-End behavior, and y-intercept. Then sketch it.	
Factor it	Suppose we are unable to factor it. Use the Rational Root Theorem to help.
Find possible zeros	Factors of lead coefficient: $\pm 1, \pm 3$ Factors of last term: $\pm 1, \pm 2$ Possible zeros: $\pm\frac{1}{1}, \pm\frac{2}{1}, \pm\frac{1}{3}, \pm\frac{2}{3}$ Possible zeros: $\pm 1, \pm 2, \pm\frac{1}{3}, \pm\frac{2}{3}$
Note	If c is a zero of a polynomial function, then $(x - c)$ is a linear factor. When a polynomial function is divided by a linear factor, the remainder $= 0$. Possible Linear Factors: $(x \pm 1), (x \pm 2), \left(x \pm \frac{1}{3}\right), \left(x \pm \frac{2}{3}\right)$
Find a factor	Use Synthetic Division to divide $f(x)$ by the possible linear factors. If the remainder is zero, then it is a factor!

	Rational Roots -- Ex. 1b
\multicolumn{2}{l	}{Given: $f(x) = 3x^2 - 5x - 2$}
\multicolumn{2}{l	}{Find: Degree, Zeros (x intercepts), Multiplicity, Far-End behavior, and y-intercept. Then sketch it.}
Previously Found	Possible Linear Factors: $(x \pm 1), (x \pm 2), \left(x \pm \frac{1}{3}\right), \left(x \pm \frac{2}{3}\right)$
Synthetic Division Check $(x - 1)$	$\begin{array}{c\|ccc} 1 & 3 & -5 & -2 \\ & \downarrow & 3 & -2 \\ \hline & 3 & -2 & -4 \end{array}$ Remainder $\neq 0$ ☹
Synthetic Division Check $(x + 1)$	$\begin{array}{c\|ccc} -1 & 3 & -5 & -2 \\ & \downarrow & -3 & 8 \\ \hline & 3 & -8 & 6 \end{array}$ Remainder $\neq 0$ ☹
Synthetic Division Check $(x - 2)$	$\begin{array}{c\|ccc} 2 & 3 & -5 & -2 \\ & \downarrow & 6 & 2 \\ \hline & 3 & 1 & 0 \end{array}$ Remainder $= 0$ ☺ Success!!! $f(x) = (x - 2) \cdot (3x + 1)$ $f(x) = (x - 2)\left(\frac{1}{3}\right)\left(x + \frac{1}{3}\right)$

Rational Roots -- Ex. 1c	
Given: $f(x) = 3x^2 - 5x - 2$ Find: Degree, Zeros (x intercepts), Multiplicity, Far-End behavior, and y-intercept. Then sketch it.	
Previously Found	$f(x) = \frac{1}{3}(x-2)\left(x+\frac{1}{3}\right)$
Degree	$D = 1 + 1 = 2$ (even)
Zeros and Multiplicity	$-\frac{1}{3}$, $m1$ 2, $m1$ Thru Thru
Far-End Behavior	Positive leading coefficient and even degree. ↖ ↗
y-intercept	y-intercept occurs at $x = 0$ $f(0) = -2$ Point: $(0, -2)$
Graph	

	Irrational Roots -- Ex. 2
	Given: $f(x) = x^3 + 5x^2 - 3x - 15$ With: One zero at $x = \sqrt{3}$ Find: All zeros and write the function in factored form with all linear (1ˢᵗ degree) factors.
Two zeros	$x = \sqrt{3}$ and $x = -\sqrt{3}$
Linear Factors	$(x - \sqrt{3})$ and $(x + \sqrt{3})$
Multiply	$(x - \sqrt{3})(x + \sqrt{3}) = x^2 - 3$
Long Division	$\begin{array}{r} x + 5 \\ x^2 + 0x - 3 \overline{\smash{)}x^3 + 5x^2 - 3x - 15} \\ -(x^3 + 0x^2 - 3x) \\ \hline 5x^2 + 0x - 15 \\ -(5x^2 + 0x - 15) \\ \hline 0 \end{array}$
Factored Form	$f(x) = x^3 + 5x^2 - 3x - 15$ $f(x) = (x^2 - 3)(x + 5)$ $f(x) = (x - \sqrt{3})(x + \sqrt{3})(x + 5)$

	Imaginary Roots -- Ex. 3
	Given: $f(x) = x^3 - 5x^2 + 8x - 6$ With: One zero at $x = 1 + i$ Find: All zeros and write the function in factored form with all linear (1st degree) factors.
Two zeros	$x = (1 + i)$ and $x = (1 - i)$
Linear Factors	$(x - (1 + i)) = (x - 1 - i)$ $(x - (1 - i)) = (x - 1 + i)$
Multiply	$(x - 1 - i)(x - 1 + i) = x^2 - 2x + 2$
Long Division	$$\begin{array}{r} x - 3 \\ x^2 - 2x + 2 \overline{\smash{\big)}\, x^3 - 5x^2 + 8x - 6} \\ -(x^3 - 2x^2 + 2x) \\ \hline -3x^2 + 6x - 6 \\ -(-3x^2 + 6x - 6) \\ \hline 0 \end{array}$$
Factored Form	$f(x) = x^3 - 5x^2 + 8x - 6$ $f(x) = (x^2 - 2x + 2\ )(x - 3)$ $= (x - (1 + i))\,(x - (1 - i))\,(x - 3)$

Infinite Possibilities -- Ex. 4a

Each root is associated with a linear factor.
If "c" is a root, then $(x - c)$ is a linear factor.

In a previous problem, we found this:
$f(x) = x^3 + 5x^2 - 3x - 15$
$f(x) = (x^2 - 3)(x + 5)$
$f(x) = (x - \sqrt{3})(x + \sqrt{3})(x + 5)$

But, there are an infinite number of functions that have the associated zeros: $-5, -\sqrt{3}, \sqrt{3}$

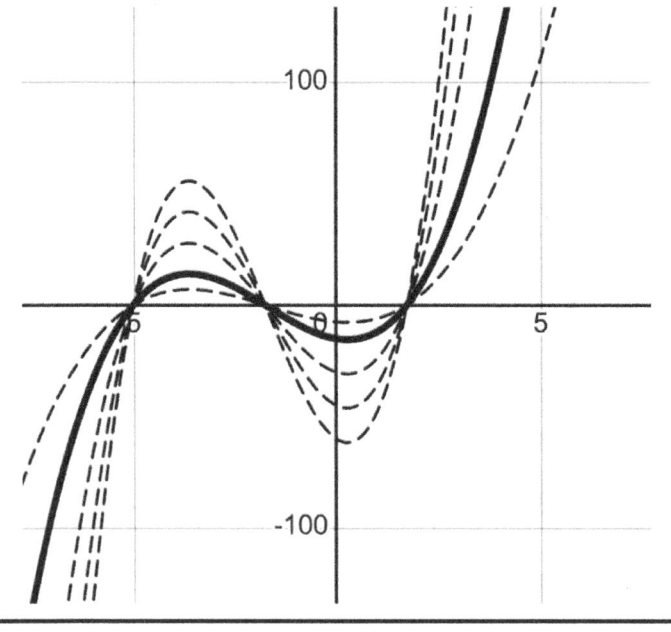

Infinite Possibilities -- Ex. 4b
Given: $f(x) = a(x - \sqrt{3})(x + \sqrt{3})(x + 5)$ Find the value of "a" (Vertical stretch)   that makes the function go through point $(4, 150)$

Insert the given point: $(x, y) = (4, 150)$ into the equation and solve for "a."
$f(x) = a(x - \sqrt{3})(x + \sqrt{3})(x + 5)$ $150 = a(4 - \sqrt{3})(4 + \sqrt{3})(4 + 5)$ $150 = a(117)$ $a = \dfrac{150}{117} = \dfrac{50}{39}$
The function, with zeros at: $-5, -\sqrt{3}, \sqrt{3}$ That includes the point: $(4, 150)$ is: $f(x) = \left(\dfrac{50}{39}\right)(x - \sqrt{3})(x + \sqrt{3})(x + 5)$
Graph on next page.

Infinite Possibilities -- Ex. 4c

Given: $f(x) = a(x - \sqrt{3})(x + \sqrt{3})(x + 5)$

Find the value of "a" (Vertical stretch)

that makes the function go through point $(4, 150)$

Previously found ...

The function, with zeros at: $-5, -\sqrt{3}, \sqrt{3}$

That includes the point: $(4, 150)$ is:

$f(x) = \left(\dfrac{50}{39}\right)(x - \sqrt{3})(x + \sqrt{3})(x + 5)$

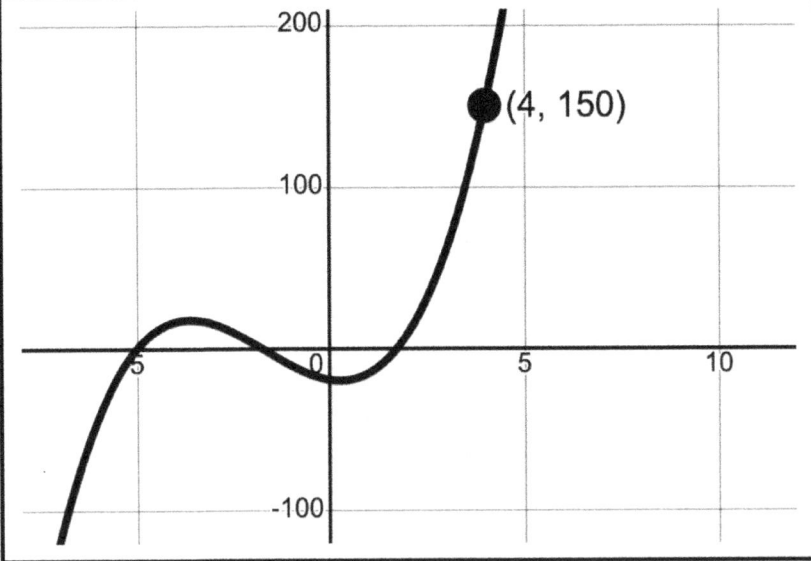

Binomial Expansion

Binomial Expansion

Binomial Expansion is finding the **coefficients** of the polynomial, when expanding a binomial, raised to a power, e.g. $(a + b)^n$.

The coefficients of the expanded polynomial can be found with the binomial expansion formula or with Pascal's Triangle. Both, shown below.

Perfect Sq. Trinomials (Review)	$(a + b)^2 = a^2 + 2ab + b^2$ $(a - b)^2 = a^2 - 2ab + b^2$
Binomial Expansion	$(a + b)^n = \sum_{k=0}^{n} \binom{n}{k} a^{n-k} \cdot b^k$
Combination	$\binom{n}{k} = \dfrac{n!}{k! \cdot (n-k)!}$
Pascal's Triangle	$\begin{array}{ccccccccccc} & & & & & 1 & & & & & \\ & & & & 1 & & 1 & & & & \\ & & & 1 & & 2 & & 1 & & & \\ & & 1 & & 3 & & 3 & & 1 & & \\ & 1 & & 4 & & 6 & & 4 & & 1 & \\ 1 & & 5 & & 10 & & 10 & & 5 & & 1 \end{array}$

Binomial Expansion – Ex. 1a
Expand $(x + 5)^3$ in two ways. First with Pascal's Triangle. Then, use the binomial expansion formula.

Binomial Expansion With Pascal's Triangle	
How many terms and coefficients?	The degree of the expanded polynomial will be $n = 3$. There will be $(n + 1) = 4$ terms. Each term has a coefficient.
Find the 4 coefficients Using Pascal's Triangle.	Find a row with 4 coefficients. The coefficients are: $1, 3, 3, 1$

$(x + 5)^3 =$
$= 1(x^3 5^0) + 3(x^2 5^1) + 3(x^1 5^2) + 1(x^0 5^3)$
$= x^3 + 15x^2 + 75x + 125$

Notes:
- The exponents decrease on x & increase on 5
- Sum of exponents $= n = 3$

Binomial Expansion – Ex. 1b

Expand $(x+5)^3$ in two ways.
First with Pascal's Triangle.
Then, use the binomial expansion formula.

Binomial Expansion With Expansion Formula

Binomial Expansion	$(a+b)^n = \sum_{k=0}^{n} \binom{n}{k} a^{n-k} \cdot b^k$
Combination	$\binom{n}{k} = \dfrac{n!}{k! \cdot (n-k)!}$
Combination on TI Calculator	Press [math], PROB, $_nC_r$ Then, fill in the values for n & r

$(x+5)^3 = \sum_{k=0}^{3} \binom{3}{k} a^{3-k} \cdot b^k$

$= \binom{3}{0} x^3 5^0 + \binom{3}{1} x^2 5^1 + \binom{3}{2} x^1 5^2 + \binom{3}{3} x^0 5^3$

$= (1)x^3 5^0 + (3)x^2 5^1 + (3)x^1 5^2 + (1)x^0 5^3$

$= (1)x^3(1) + (3)x^2(5) + (3)(x)5^2 + (1)(1)5^3$

$= x^3 + 15x^2 + 75(x) + 125$

Binomial Expansion – Ex. 2

Expand $(a + b)^4$ using the binomial expansion formula.

Binomial Expansion	$(a + b)^n = \sum_{k=0}^{n} \binom{n}{k} a^{n-k} \cdot b^k$

Binomial Expansion With Expansion Formula

$$(a + b)^4 = \sum_{k=0}^{4} \binom{4}{k} a^{4-k} \cdot b^k$$

$$= \binom{4}{0} a^4 b^0 + \binom{4}{1} a^3 b^1 + \binom{4}{2} a^2 b^2$$
$$+ \binom{4}{3} a^1 b^3 + \binom{4}{4} a^0 b^4$$

$$= (1)a^4 + (4)a^3 b^1 + (6)a^2 b^2$$
$$+ (4)a^1 b^3 + (1)b^4$$

$$= a^4 + 4a^3 b + 6a^2 b^2 + 4ab^3 + b^4$$

Binomial Expansion – Ex. 3

Expand $(x + 2)^3$ and then expand $(x - 2)^3$ using the binomial expansion formula.

Binomial Expansion	$(a + b)^n = \sum_{k=0}^{n} \binom{n}{k} a^{n-k} \cdot b^k$

$(x + 2)^3 = \sum_{k=0}^{3} \binom{3}{k} x^{3-k} \cdot 2^k$

$= \binom{3}{0} x^3 2^0 + \binom{3}{1} x^2 2^1 + \binom{3}{2} x^1 2^2 + \binom{3}{3} x^0 2^3$

$= (1)x^3 2^0 + (3)x^2 2^1 + (3)x^1 2^2 + (1)x^0 2^3$

$= x^3 + 6x^2 + 12x + 8$

$(x - 2)^3 = (x + (-2))^3$

$= \sum_{k=0}^{3} \binom{3}{k} x^{3-k} \cdot (-2)^k$

$= (1)x^3(-2)^0 + (3)x^2(-2)^1 + (3)x^1(-2)^2 + (1)x^0(-2)^3$

$= x^3 - 6x^2 + 12x - 8$

Binomial Expansion – Ex. 4

Write the 6th term of $(x + y)^8$

Binomial Expansion	$(x + y)^8 = \sum_{k=0}^{8} \binom{8}{k} x^{8-k} \cdot y^k$
Notes	<ul><li>Total of 9 terms</li><li>k starts counting at 0</li><li>6th term has $k = 5$</li><li>6th term is: $\binom{8}{5} x^{8-5} \cdot y^5$</li></ul>
6th Term	$\binom{8}{5} x^{8-5} \cdot y^5 = (56) x^3 y^5$

Binomial Expansion – Ex. 5		
Which of the following terms is included in the expansion of: $(g+h)^{10}$?		
[A] $10\,g^5h^5$		[B] $120\,g^5h^5$
[C] $210\,g^5h^5$		[D] $252\,g^5h^5$

Binomial Expansion	$(g+h)^{10} = \sum_{k=0}^{10} \binom{10}{k} g^{10-k} \cdot h^k$
Notes	• Total of 11 terms • k starts counting at 0 • Exponents on both variables is 5. Therefore: $k = 5$ • The 6th term has $k = 5$ • Calculate the coefficient of 6th term. $\binom{10}{k} = \binom{10}{5} = 252$
Answer	The answer is [D]

Partial Fractions

Partial Fraction Decomposition
With **Partial Fraction Decomposition**, a rational expression is rewritten as the sum of simple fractions. For example: $\dfrac{x+7}{x^2-x-6} = \dfrac{2}{x-3} - \dfrac{1}{x+2}$

	$\dfrac{Numerator}{Denominator} = \dfrac{N(x)}{D(x)} = Decomposition$
1	If Improper Fraction, divide numerator by the denominator.
2	Factor denominator into linear factors or irreducible 2nd degree (quadratic) factors.
3	For each linear factor, with multiplicity m, of the form $(ax+b)^m$ the decomposition must include: $\dfrac{A_1}{(ax+b)} + \dfrac{A_2}{(ax+b)^2} + \cdots + \dfrac{A_m}{(ax+b)^m}$
4	For each quadratic factor, with multiplicity n, of the form $(ax^2+bx+c)^n$ the decomposition must include: $\dfrac{B_1 x + C_1}{(ax^2+bx+c)^1} + \cdots + \dfrac{B_n x + C_n}{(ax^2+bx+c)^n}$

Partial Fraction Decomposition
Distinct Linear Factors – Ex. 1

Write the partial fraction decomposition for

$$\frac{x+7}{x^2-x-6}$$

Not an improper fraction.	No division needed.
Factor denom.	$x^2 - x - 6 = (x-3)(x+2)$
Break Down	$\dfrac{x+7}{(x-3)(x+2)} = \dfrac{A}{(x-3)} + \dfrac{B}{(x+2)}$ $x + 7 = A(x+2) + B(x-3)$
	$x + 7 = A(x+2) + B(x-3)$
Solve for A & B	$x = -2 \rightarrow \quad 5 = -5B$ $\quad\quad\quad\quad\quad\quad B = -1$
	$x = 3 \rightarrow \quad 10 = 5A$ $\quad\quad\quad\quad\quad\quad A = 2$
	$\dfrac{x+7}{x^2-x-6} = \dfrac{2}{(x-3)} - \dfrac{1}{(x+2)}$

Partial Fraction Decomposition
Repeated Linear Factors – Ex. 2a

Write the partial fraction decomposition for

$$\frac{5x^2 + 20x + 6}{x^3 + 2x^2 + x}$$

$$x^3 + 2x^2 + x = x(x^2 + 2x + 1)$$
$$= x(x + 1)(x + 1)$$
$$= x(x + 1)^2$$

$$\frac{5x^2 + 20x + 6}{x(x + 1)^2} = \frac{A}{x} + \frac{B}{(x + 1)} + \frac{C}{(x + 1)^2}$$

$$5x^2 + 20x + 6 = A(x + 1)^2 + Bx(x + 1) + Cx$$

$x = 0 \quad \rightarrow \quad 6 = A$

$x = -1 \quad \rightarrow \quad -9 = -C$
$$C = 9$$

$x = 1 \quad \rightarrow \quad 31 = 4A + 2B + C$
$$31 = 4(6) + 2B + 9$$
$$B = -1$$

$$\frac{5x^2 + 20x + 6}{x^3 + 2x^2 + x} = \frac{6}{x} - \frac{1}{(x + 1)} + \frac{9}{(x + 1)^2}$$

Partial Fraction Decomposition
Repeated Linear Factors – Ex. 2b

Write the partial fraction decomposition for

$$\frac{5x^2 + 20x + 6}{x^3 + 2x^2 + x}$$

$$x^3 + 2x^2 + x = x(x^2 + 2x + 1)$$
$$= x(x+1)(x+1)$$
$$= x(x+1)^2$$

$$\frac{5x^2 + 20x + 6}{x(x+1)^2} = \frac{A}{x} + \frac{B}{(x+1)} + \frac{C}{(x+1)^2}$$

$$5x^2 + 20x + 6 = A(x+1)^2 + Bx(x+1) + Cx$$
$$= A(x^2 + 2x + 1) + B(x^2 + x) + Cx$$
$$= x^2(A+B) + x(2A+B+C) + (A)$$

Match coefficients of: x^2, x^1, x^0

$5 = A + B$	$5 = 6 + B \rightarrow B = -1$
$20 = 2A + B + C$	$C = 20 - 2A - B$
$6 = A$	$C = 20 - 2(6) + 1$
	$C = 9$

$$\frac{5x^2 + 20x + 6}{x^3 + 2x^2 + x} = \frac{6}{x} - \frac{1}{(x+1)} + \frac{9}{(x+1)^2}$$

Partial Fraction Decomposition
Distinct Linear & Quadratic Factors — Ex. 3

Write the partial fraction decomposition for

$$\frac{x^2 + 4x + 4}{x^3 - x^2 + 2x - 2}$$

$x^3 - x^2 + 2x - 2 = (x-1)(x^2 + 2)$

For help with factoring, see the "Roots" section in this book. If "c" is a zero, then $(x - c)$ is a linear factor.

$$\frac{x^2 + 4x + 4}{(x-1)(x^2+2)} = \frac{A}{(x-1)} + \frac{Bx + C}{(x^2 + 2)}$$

$$x^2 + 4x + 4 = A(x^2 + 2) + (Bx + C)(x - 1)$$

$$= x^2(A + B) + x(-B + C) + (2A - C)$$

Match coefficients of: x^2, x^1, x^0

	Solve for A, B, C	
$1 = A + B$		$A = 3$
$4 = -B + C$		$B = -2$
$4 = (2A - C)$		$C = 2$

$$\frac{x^2 + 4x + 4}{(x-1)(x^2+2)} = \frac{3}{(x-1)} + \frac{-2x + 2}{(x^2 + 2)}$$

Partial Fraction Decomposition
Repeated Quadratic Factors – Ex. 4

Write the partial fraction decomposition for

$$\frac{8x^3 + 13x}{(x^2 + 2)^2}$$

Here, the denominator is already factored.

$$\frac{8x^3 + 13x}{(x^2 + 2)^2} = \frac{Ax + B}{(x^2 + 2)} + \frac{Cx + D}{(x^2 + 2)^2}$$

$$8x^3 + 13x = (Ax + B)(x^2 + 2) + (Cx + D)$$
$$= x^3(A) + x^2(B) + x(2A + C) + (2B + D)$$

Match coefficients of: x^3, x^2, x^1, x^0

$8 = A$		$A = 8$
$0 = B$	Solve for	$B = 0$
$13 = (2A + C)$	A, B, C, D	$C = -3$
$0 = 2B + D$		$D = 0$

$$\frac{8x^3 + 13x}{(x^2 + 2)^2} = \frac{8x}{(x^2 + 2)} - \frac{3x}{(x^2 + 2)^2}$$

Partial Fraction Decomposition
Solving for the Coefficients – Ex. 5a

Partial Fraction Decomposition is a useful technique for simplifying complex fractions into simple fractions. The last step of this technique is solving for the coefficients. Often, this is done algebraically.

Another way to solve a system of equations is with matrices. Matrices is a topic covered in another section, but will be used here to demonstrate how to use matrices to solve a system of equations.

For the previous example, the following system of equations was solved, using algebra. On the next page, the same system will be solved using matrices.

$8 = A$		$A = 8$
$0 = B$	Solve for	$B = 0$
$13 = (2A + C)$	A, B, C, D	$C = -3$
$0 = 2B + D$		$D = 0$

Partial Fraction Decomposition Solving for the Coefficients – Ex. 5b	
Solve the system of equations on a TI-84 calculator.	$8 = A$ $0 = B$ $13 = (2A + C)$ $0 = 2B + D$

Put in matrix form	$A = 8$ $B = 0$ $2A + C = 13$ $2B + D = 0$	$\begin{array}{cccc|c} A & B & C & D & = n \\ 1 & 0 & 0 & 0 & 8 \\ 0 & 1 & 0 & 0 & 0 \\ 2 & 0 & 1 & 0 & 13 \\ 0 & 2 & 0 & 1 & 0 \end{array}$

Put matrix in TI-84 calculator and solve.	
[2nd] matrix, EDIT, [enter] Size 4 X 5 (rows & cols.)	MATRIX[A] 4 ×5 $\begin{bmatrix} 1 & 0 & 0 & 0 & 8 \\ 0 & 1 & 0 & 0 & 0 \\ 2 & 0 & 1 & 0 & 13 \\ 0 & 2 & 0 & 1 & 0 \end{bmatrix}$
[2nd] matrix, MATH, rref, [enter]	NAMES **MATH** EDIT 8↑Matr▶list(9:List▶matr(0:cumSum(A:ref(**B:**rref(

Partial Fraction Decomposition
Solving for the Coefficients — Ex. 5c

	Put matrix in TI-84 calculator and solve.
[2nd] matrix, EDIT, [enter] Size 4 X 5 (rows & cols.)	MATRIX[A] 4 ×5 $\begin{bmatrix} 1 & 0 & 0 & 0 & 8 \\ 0 & 1 & 0 & 0 & 0 \\ 2 & 0 & 1 & 0 & 13 \\ 0 & 2 & 0 & 1 & 0 \end{bmatrix}$
[2nd] matrix, MATH, rref, [enter]	NAMES **MATH** EDIT 8↑Matr▶list(9:List▶matr(0:cumSum(A:ref(**B:**rref(
[2nd] matrix, NAMES, [A], [enter]	rref([A] rref([A]) Close parenthesis.
Answers in last column.	rref([A]) $\begin{bmatrix} 1 & 0 & 0 & 0 & 8 \\ 0 & 1 & 0 & 0 & 0 \\ 0 & 0 & 1 & 0 & -3 \\ 0 & 0 & 0 & 1 & 0 \end{bmatrix}$
Answers	$(A, B, C, D) = (8, 0, -3, 0)$

Partial Fraction Decomposition
Solving for the Coefficients — Ex. 5d

	Put matrix in Desmos and solve. www.desmos.com/matrix
New Matrix 4 Rows 5 Cols Enter	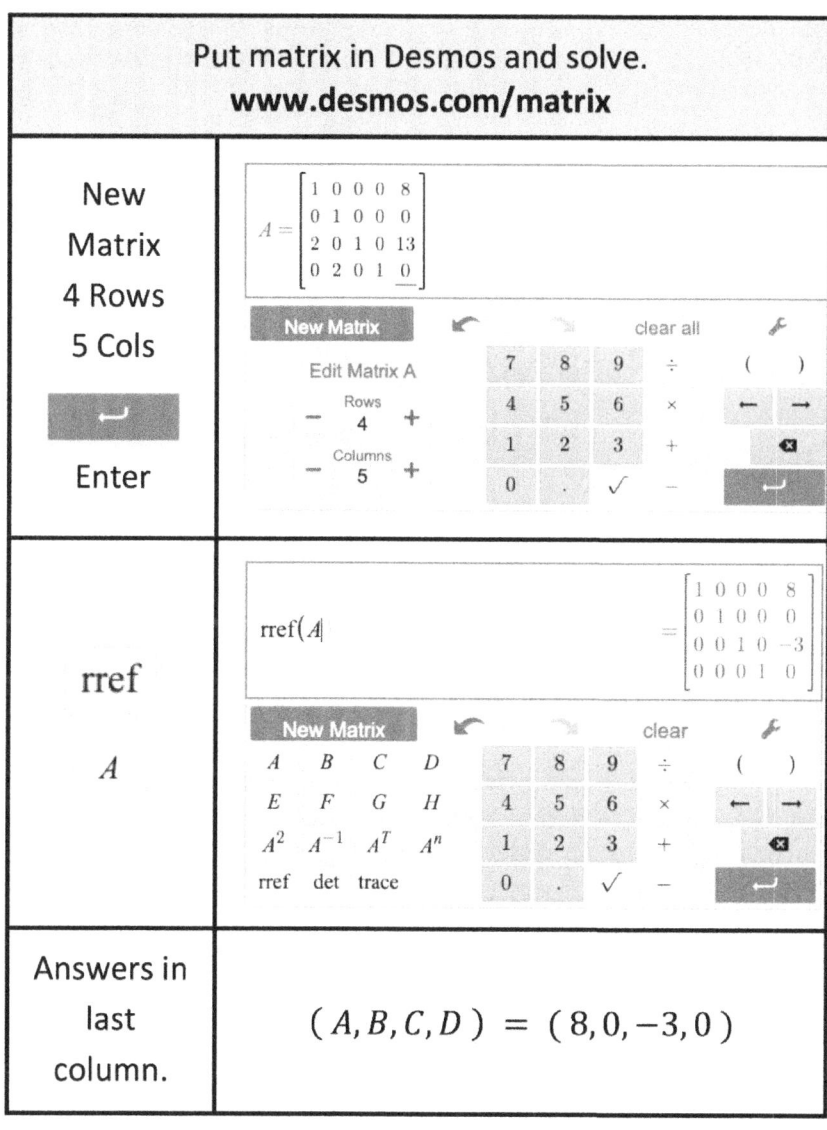
rref A	
Answers in last column.	$(A, B, C, D) = (8, 0, -3, 0)$

Conics

Parabola -- Review

The following second-degree equations (quadratic equations) can be used to represent parabolas.

Standard Form	$y = ax^2 + bx + c$
Vertex Form	$y = a(x - h)^2 + k$
Factored Form	$y = a(x - z_1)(x - z_2)$

Conic Section
Another Way to Represent a Parabola

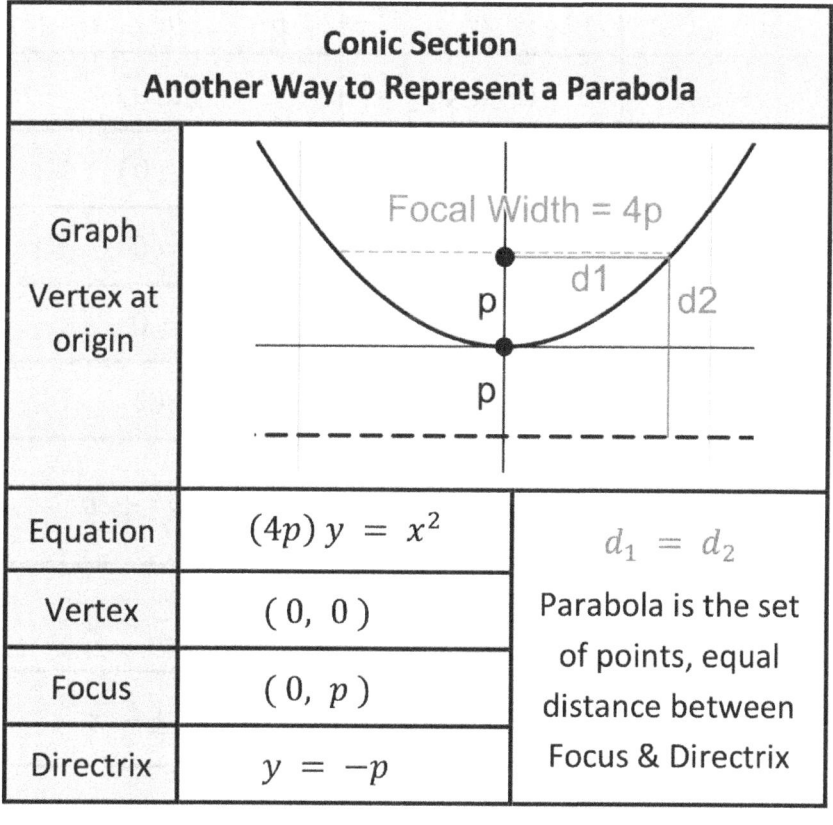

Graph Vertex at origin		
Equation	$(4p)y = x^2$	$d_1 = d_2$
Vertex	(0, 0)	Parabola is the set of points, equal distance between Focus & Directrix
Focus	(0, p)	
Directrix	$y = -p$	

Conic Sections – Ellipse & Hyperbola

	Ellipse	Hyperbola
Graph		
Equation	$\dfrac{x^2}{a^2} + \dfrac{y^2}{b^2} = 1$	$\dfrac{x^2}{a^2} - \dfrac{y^2}{b^2} = 1$
Vertex	$(\pm a, 0)$	$(\pm a, 0)$
Foci	$(\pm c, 0)$	$(\pm c, 0)$
Major Axis	$y = 0$	$y = 0$
Minor Axis	$x = 0$	$x = 0$
Center	$(0, 0)$	$(0, 0)$
c	$c^2 = a^2 - b^2$ $a > b$	$c^2 = a^2 + b^2$
Eccentricity	$e = \dfrac{c}{a} < 1$	$e = \dfrac{c}{a} > 1$
Asymptote	---	$y = \pm \dfrac{b}{a} x$

Conics -- Ex. 1	
Given: $f(x) = \left(\frac{1}{8}\right)x^2 + 4$ Convert it to a conic section form. Find coordinates of the vertex, focus, and equation for directrix. Sketch it.	
Rearrange To get $(4p)y = x^2$	$y = \left(\frac{1}{8}\right)x^2 + 4$ $y - 4 = \left(\frac{1}{8}\right)x^2$ Opens up $8(y - 4) = x^2$ V. Shift up 4
Vertex	$(x, y) = (h, k) = (0, 4)$
Find "p"	$4p = 8$ $p = 2$
Focus	$(h, k + p) = (0, 4 + 2) = (0, 6)$
Directrix	$y = k - p$ $y = 4 - 2$ $y = 2$
Sketch	

Conics -- Ex. 2a

Given: $4x^2 + 9y^2 - 48x + 72y + 144 = 0$

Convert it to a conic section form of an ellipse.
Find coordinates of the center and foci.
Hint: Complete the square twice. Sketch.

$4x^2 + 9y^2 - 48x + 72y + 144 = 0$

$(4x^2 - 48x) + (9y^2 + 72y) = -144$

$4(x^2 - 12x) + 9(y^2 + 8y) = -144$

$4(x^2 - 12x + 36) + 9(y^2 + 8y + 16)$
$= -144 + 4(36) + 9(16)$

$4(x-6)^2 + 9(y+4)^2 = 144$

$$\frac{4(x-6)^2}{144} + \frac{9(y+4)^2}{144} = \frac{144}{144}$$

$$\frac{(x-6)^2}{36} + \frac{(y+4)^2}{16} = \frac{(x-6)^2}{6^2} + \frac{(y+4)^2}{4^2} = 1$$

Center	$(h, k) = (6, -4)$
a, b	$a = 6 \quad b = 4 \quad$ Major Axis: x
c	$c = \sqrt{6^2 - 4^2} = \sqrt{20} = 2\sqrt{5}$
Foci	$(6 \pm c, -4) = (6 \pm 2\sqrt{5}, -4)$
Major Vertices	$(h \pm a, k) = (6 \pm 6, -4)$

	Conics -- Ex. 2b
	Given: $4x^2 + 9y^2 - 48x + 72y + 144 = 0$ Convert it to a conic section form of an ellipse. Find coordinates of the center and foci. Hint: Complete the square twice. Sketch.
Previously Found	$\dfrac{(x-6)^2}{6^2} + \dfrac{(y+4)^2}{4^2} = 1$ Center: $(h, k) = (6, -4)$ Foci: $\left(6 \pm 2\sqrt{5},\ -4\right)$ Major Vertices: $(6 \pm 6,\ -4)$
Sketch	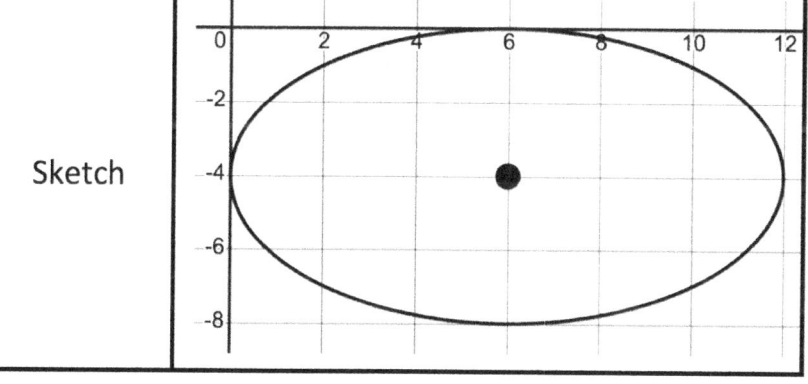

Conics -- Ex. 3
An oval, centered at the origin has $r_1 = 4$ & $r_2 = 3$
Find: The equation of the oval as an Ellipse Conic Section. 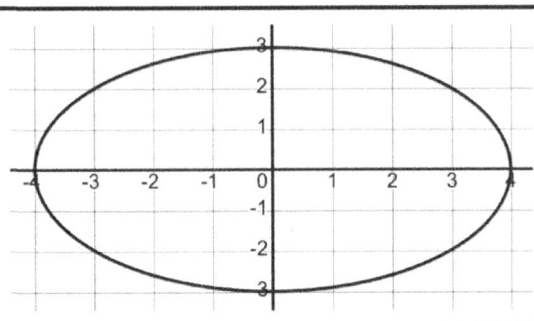

Equation	$\dfrac{x^2}{4^2} - \dfrac{y^2}{3^2} = 1$
Center	$(h, k) = (0, 0)$
a, b	$a = 4 \quad b = 3 \quad$ Major Axis: x
c	$c = \sqrt{4^2 - 3^2} = \sqrt{7}$
Foci	$(h \pm c, k) = (\pm\sqrt{7}, 0)$
Major Vertices	$(\pm 6, 0)$

	Conics -- Ex. 4		
Given this graph of a hyperbola, find the equation. Also, find eqn. of asymptotes			
Center	$(h, k) = (0, 0)$		
Major Axis	x	Because it opens on the x axis. Also, foci are on x axis.	
a, b, c	$a = 2, \; b = 4, \; c = \sqrt{a^2 + b^2} = \sqrt{20}$		
Foci	$(x, y) = (\pm\sqrt{20}, 0)$		
Equation	$\dfrac{x^2}{2^2} - \dfrac{y^2}{4^2} = 1$		Term with "x" is first because it is on the major axis.
Asymptotes	$y = \pm \dfrac{4}{2} x = \pm 2x$		

	Conics -- Ex. 5	
Given this graph of an ellipse, find the eqn.		
Center	$(h, k) = (-1, 3)$	
Major Axis	y Because it is longer on y axis. Also, foci are on y axis.	
a, b, c	$a = 4, \quad b = 2$ $c = \sqrt{a^2 - b^2} = \sqrt{12} = 2\sqrt{3}$	
Foci	$(x, y) = (-1, \; 3 \pm 2\sqrt{3})$	
Eqn.	$\dfrac{(x+1)^2}{2^2} + \dfrac{(y-3)^2}{4^2} = 1$	y^2 has larger denominator so y is major axis.

Vectors

	Vectors →	
Definition	Vector from point $A = (x_1, y_1, z_1)$ To point $B = (x_2, y_2, z_2)$ $\overrightarrow{AB} = \langle x_2 - x_1, y_2 - y_1, z_2 - z_1 \rangle$ A vector has magnitude and direction.	
Magnitude	$\vec{p} = \langle a, b, c \rangle = a\mathbf{i} + b\mathbf{j} + c\mathbf{k}$ $|\vec{p}| =$ Magnitude of $\vec{p}$ $|\vec{p}| = \sqrt{a^2 + b^2 + c^2}$	
Unit Vector	$\dfrac{\vec{p}}{|\vec{p}|} = \dfrac{\langle a,b,c \rangle}{\sqrt{a^2 + b^2 + c^2}}$	Unit vector in direction of $\vec{p}$
	$\mathbf{i}, \mathbf{j}, \mathbf{k} =$ Unit Vectors In x, y, z directions.	
Addition	$\vec{p} + \vec{q} = \langle a, b, c \rangle + \langle d, e, f \rangle$ $\vec{p} + \vec{q} = \langle a+d, b+e, c+f \rangle$	
	[diagram: triangle showing $\vec{p}$, $\vec{q}$ tip to tail with resultant $\vec{p} + \vec{q}$]	Add vectors "tip to tail"

Vectors
Dot Product and Cross Product

Dot Product	$\vec{p} \cdot \vec{q} = \langle a,b,c \rangle \cdot \langle d,e,f \rangle$ $\vec{p} \cdot \vec{q} = ad + be + cf$ (a scalar)	
	$\vec{p} \cdot \vec{q} = \|\vec{p}\|\|\vec{q}\| \cos\theta$ $\cos\theta = \dfrac{\vec{p} \cdot \vec{q}}{\|\vec{p}\|\|\vec{q}\|}$	θ = angle between $\vec{p}$ and $\vec{q}$
Cross Product	$\vec{p} \times \vec{q} = \langle a,b,c \rangle \times \langle d,e,f \rangle$ $= \begin{vmatrix} i & j & k \\ a & b & c \\ d & e & f \end{vmatrix}$ $= \begin{vmatrix} b & c \\ e & f \end{vmatrix} i - \begin{vmatrix} a & c \\ d & f \end{vmatrix} j + \begin{vmatrix} a & b \\ d & e \end{vmatrix} k$ $= \langle (bf-ce), (af-cd), (ae-bd) \rangle$ $=$ Vector $\perp$ to both $\vec{p}$ and $\vec{q}$	
	$\|\vec{p} \times \vec{q}\| = \|\vec{p}\|\|\vec{q}\| \sin\theta$	If $\vec{p} \times \vec{q} = 0$ Then $\vec{p} \parallel \vec{q}$
Volume of Parallel-epiped	$V = \|\vec{p} \cdot (\vec{q} \times \vec{v})\|$ Sides: $\vec{p}, \vec{q}, \vec{v}$	

Vectors -- Ex. 1
Given Points: $A\,(1,2)$, $B\,(7,10)$, $C\,(-5,20)$
Find Vectors: $\vec{AB}, \vec{AC}, \vec{BC}, \vec{CB}$ and their magnitudes.
Also find vector $\vec{p} = \vec{AB} - \vec{BC}$ and its magnitude.

Vector	Magnitude
$\vec{AB} = \langle 7-1, 10-2 \rangle$ $\vec{AB} = \langle 6, 8 \rangle$	$\|\vec{AB}\| = \sqrt{6^2 + 8^2}$ $\|\vec{AB}\| = \sqrt{100} = 10$
$\vec{AC} = \langle -5-1, 20-2 \rangle$ $\vec{AC} = \langle -6, 18 \rangle$	$\|\vec{AC}\| = \sqrt{6^2 + 18^2}$ $\|\vec{AC}\| = \sqrt{360} = 6\sqrt{10}$
$\vec{BC} = \langle -5-7, 20-10 \rangle$ $\vec{BC} = \langle -12, 10 \rangle$	$\|\vec{BC}\| = \sqrt{12^2 + 10^2}$ $\|\vec{BC}\| = \sqrt{244} = 2\sqrt{61}$
$\vec{CB} = \langle 7+5, 10-20 \rangle$ $\vec{CB} = \langle 12, -10 \rangle$	$\|\vec{CB}\| = \sqrt{12^2 + 10^2}$ $\|\vec{CB}\| = \sqrt{244} = 2\sqrt{61}$
$\vec{p} = \vec{AB} - \vec{BC}$ $\vec{p} = \langle 6, 8 \rangle - \langle -12, 10 \rangle$ $\vec{p} = \langle 6+12, 8-10 \rangle$ $\vec{p} = \langle 18, -2 \rangle$	$\|\vec{p}\| = \sqrt{18^2 + 2^2}$ $\|\vec{p}\| = \sqrt{328} = 2\sqrt{82}$

Vectors -- Ex. 2	
Given 3 Points: $A\,(1,2,3)$, $B\,(5,10,15)$, $C\,(11,2,0)$ Find Vector $\vec{n}$, perpendicular to the plane defined by the three given points.	
Find two vectors on the plane, with the 3 points.	$\overrightarrow{AB} = \langle 5-1, 10-2, 15-3 \rangle$ $\overrightarrow{AB} = \langle 4, 8, 12 \rangle$ $\overrightarrow{AC} = \langle 11-1, 2-2, 0-3 \rangle$ $\overrightarrow{AC} = \langle 10, 0, -3 \rangle$
Use Cross Product to find normal vector, $\vec{n}$	$\vec{n} = \overrightarrow{AB} \times \overrightarrow{AC}$ $\vec{n} = \begin{vmatrix} i & j & k \\ 4 & 8 & 12 \\ 10 & 0 & -3 \end{vmatrix}$ $= i \begin{vmatrix} 8 & 12 \\ 0 & -3 \end{vmatrix} - j \begin{vmatrix} 4 & 12 \\ 10 & -3 \end{vmatrix} + k \begin{vmatrix} 4 & 8 \\ 10 & 0 \end{vmatrix}$ $= (-24)i - (-12 - 120)j + (-80)k$ $= \langle -24, 132, -80 \rangle = 4\langle -6, 33, -20 \rangle$
$\vec{n}$	$\vec{n} = 4\langle -6, 33, -20 \rangle$

Vectors -- Ex. 3

Given 2 vectors: $\vec{a} = \langle 1,2 \rangle$ and $\vec{b} = \langle 3,4 \rangle$
Find the angle, θ, between them.

Use the Dot Product	$\vec{a} \cdot \vec{b} = \|\vec{a}\|\|\vec{b}\| \cos \theta$ $\cos \theta = \dfrac{\vec{a} \cdot \vec{b}}{\|\vec{a}\|\|\vec{b}\|}$ $\cos \theta = \dfrac{\langle 1,2 \rangle \cdot \langle 3,4 \rangle}{\sqrt{1^2 + 2^2}\sqrt{3^2 + 4^2}}$ $\cos \theta = \dfrac{3 + 8}{\sqrt{5}\sqrt{25}}$ $\cos \theta = \dfrac{11}{5\sqrt{5}} \approx .98387$
	$\theta = \cos^{-1}(.98387)$ $\theta = 10.3°$ (About 10 degrees)
Sketch	(graph showing vectors $\vec{a}$ and $\vec{b}$ on coordinate plane from 0 to 3 on x-axis, 0 to 4 on y-axis)

Parametric Equations

	Plane Curves and Parametric Equations
If	f and g are continuous functions of t on an interval I,
Then	the set of ordered pairs $(f(t), g(t))$ Is a **Plane Curve** C.
The equations:	$x = f(t)$ and $y = g(t)$ are **Parametric Equations** for C, and t is the parameter

Example

Sketch the parametric curve for the following set of parametric equations: $x = t^2 - 1$ $\quad y = t + 1$

t	x	y
0	−1	1
1	0	2
2	3	3
3	8	4
4	15	5

Plot the points.

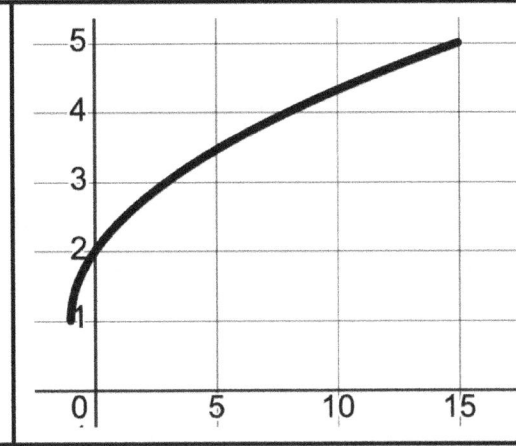

Plane Curves and Parametric Equations With Graphing Tools

Sketch the parametric curve for the following set of parametric equations: $x = t^2 - 1 \qquad y = t + 1$

Desmos

$f(t) = t^2 - 1$

$g(t) = t + 1$

$(f(t), g(t))$

$0 \leq t \leq 4$

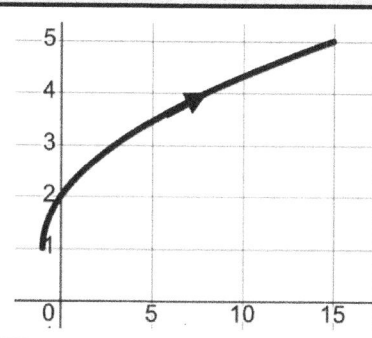

TI-84 Graphing Calculator

[mode]   Parametric
[y=]   Type eqns.
Note: [XTθn] $= T$

[window]

Plot1 Plot2
$X_{1T} = T^2 - 1$
$Y_{1T} = T + 1$

WINDOW
Tmin=■
Tmax=4
Tstep=1
Xmin=-5

[graph], [trace]

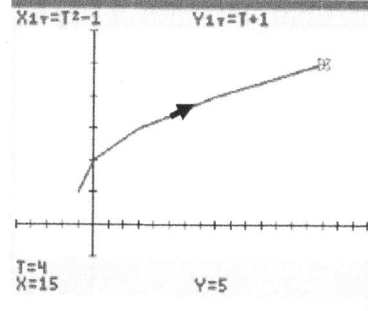

Note: Graphing tools do not include arrows. You should add them.

Eliminate the Parameter – Ex. 1

Eliminate the parameter in the parametric equations:
$$x = t^2 - 3t \qquad y = t - 1$$

Solve for t in one equation.	$y = t - 1$ $t = y + 1$
Substitute for t in the other.	$x = t^2 - 3t$ $x = (y+1)^2 - 3(y+1)$
Reformat to get $y = f(x)$	$x = (y^2 + 2y + 1) - 3y - 3$ $x = y^2 - y - 2$
Complete the square.	$y^2 - y = x + 2$ $y^2 - y + \frac{1}{4} = x + 2 + \frac{1}{4}$ $\left(y - \frac{1}{2}\right)^2 = x + \frac{9}{4}$
Solve for y	$y - \frac{1}{2} = \pm\sqrt{x + \frac{9}{4}}$ $y = \frac{1}{2} \pm \sqrt{x + \frac{9}{4}}$

Eliminate the Parameter – Ex. 2

Eliminate the parameter and graph the curve represented by the parametric equations:
$$x = \cos t \qquad y = 3 - \sin^2 t$$

Recall	$\sin^2 t + \cos^2 t = 1$
Use Trig. Identity with one equation.	$x = \cos t$ $x^2 = \cos^2 t$ $x^2 = 1 - \sin^2 t$ $\sin^2 t = 1 - x^2$
Remove "t" from the other eqn. Get: $y = f(x)$	$y = 3 - \sin^2 t$ $y = 3 - (1 - x^2)$ $y = 2 + x^2$ A parabola
[y=] [window] Plot1 Plot2 Plot3 \X₁ᴛ ▣ os(T) Y₁ᴛ ▣ 3-(sin(T))² WINDOW Tmin=0 Tmax=180 Tstep=1 [graph], [trace]	X₁ᴛ=cos(T) Y₁ᴛ=3-(sin(T))² T=30 X=0.8660254 Y=2.75 [trace] helps find direction.

Find Parametric Equation for a Curve – Ex. 3a

Find the parametric equations for the line with slope 3 that passes through the point (1,5)

Find equation	$y = mx + b$ $y = 3x + b$	Slope $= 3$
Solve for b	$(x, y) = (1,5)$ $5 = 3(1) + b$ $2 = b$	
Equation	$y = 3x + 2$	
Let $x = t$	$y = 3t + 2$	
Parametric Equations	$x = t$ $y = 3t + 2$	

Find Parametric Equation for a Curve – Ex. 3b

Find the parametric equations for the line with slope 3 that passes through the point (1,5)

A solution with vectors ...

Equation of a vector, $\vec{v}$, through the given point with a slope of 3	$\vec{v} = \langle 1,5 \rangle + t \langle 1,3 \rangle$ Note: The parameter, t, can be positive or negative.
Parametric Equations	$x = 1 + t$ $y = 5 + 3t$

Note: In the previous solution, if you let $x = 1 + t$
Then you get the same set of parametric equations.

Find Parametric Equation for a Curve – Ex. 4a

Find the parametric equations for the line through points $A(-2, 4)$ and $B(3, 6)$.

Find vector $\vec{AB}$	$\vec{AB} = \langle 3+2, 6-4 \rangle$ $\vec{AB} = \langle 5, 2 \rangle$
$O\ (0,0)$ is the origin. $P(x, y)$ is a point on the line through A & B	$\vec{OP} = \vec{OA} + \vec{AP}$ $\vec{OP} = \vec{OA} + t\vec{AB}$ $\vec{OP} = \langle -2, 4 \rangle + t \langle 5, 2 \rangle$ Note: Could have used $\vec{OB}$
Parametric Equations	$x = -2 + 5t$ $y = 4 + 2t$

$f(t) = -2 + 5t$

$g(t) = 4 + 2t$

$(f(t), g(t))$

$0 \leq t \leq 4$

$(-2, 4)$
Label

$(3, 6)$
Label

Find Parametric Equation for a Curve – Ex. 4b

Find the parametric equations for the line through points $A(-2, 4)$ and $B(3, 6)$

A different solution...

Find eqn. of a line through A and B.	$m = \dfrac{6-4}{3+2} = \dfrac{2}{5}$ $y = \left(\dfrac{2}{5}\right)x + b$
Use point (3,6) to find "b"	$6 = \left(\dfrac{2}{5}\right)(3) + b$ $b = 6 - \dfrac{6}{5} = \dfrac{24}{5}$
Equation	$y = \left(\dfrac{2}{5}\right)x + \dfrac{24}{5}$
Let $x = t$	$y = \left(\dfrac{2}{5}\right)t + \dfrac{24}{5}$
Parametric Eqns.	$x = t \qquad y = \dfrac{24}{5} + \dfrac{2}{5}t$

$f(t) = t$

$g(t) = \dfrac{24}{5} + \dfrac{2}{5}t$

$(f(t), g(t))$

$-5 \leq t \leq 5$

$(-2, 4)$
Label

$(3, 6)$

Series

Sequence and Series	
Sequence	An arrangement of numbers in a particular order.
Series	The sum of the elements of a sequence.

Arithmetic Sequence Have a common difference between elements (d)	
Recursive	$a_n = a_{n-1} + d$; $a_1 = given$
Explicit	$a_n = a_1 + d(n-1)$

Geometric Sequence Have a common ratio between elements (r)	
Recursive	$a_n = a_{n-1}(r)$; $a_1 = given$
Explicit	$a_n = a_1(r)^{n-1}$

	Arithmetic Sequence **Have a common difference between elements (d)**
Sequence Example	$1, 3, 5, 7, 9, \ldots \quad d = 2$
Series Example	$1 + 3 + 5 + 7 + 9 + \cdots$
Recursive	$a_n = a_{n-1} + d \qquad ; a_1 = given$
Explicit	$a_n = a_1 + d(n-1)$
Sum of first n terms	$S_n = \sum_{i=1}^{n} a_1 + d(i-1)$ $S_n = n \cdot (Average)$ $S_n = n \cdot \left(\dfrac{a_1 + a_n}{2} \right)$
Sum of Infinite Series	Not possible For Arithmetic Sequence.

	Geometric Sequence
	Have a common ratio between elements (r)
Sequence Example	$1, 2, 4, 8, 16, \ldots \quad r = 2$
Series Example	$1 + 2 + 4 + 8 + 16 + \cdots$
Recursive	$a_n = a_{n-1}(r) \qquad ; a_1 = given$
Explicit	$a_n = a_1(r)^{n-1}$
Sum of first n terms	$S_n = \sum_{i=1}^{n} a_1 \cdot r^{i-1}$ $S_n = a_1 \left(\dfrac{1 - r^n}{1 - r} \right)$ $S_n = \dfrac{a_1 - a_n(r)}{1 - r}$
Sum of Infinite Series	Only possible if: $\lvert r \rvert < 1$ $S_\infty = \sum_{i=1}^{\infty} a_1 \cdot r^{i-1}$ $S_\infty = a_1 \left(\dfrac{1 - r^n}{1 - r} \right) \quad \nearrow 0$ $S_\infty = \dfrac{a_1}{1 - r}$

Arithmetic Mean
Arithmetic Mean $= Average$ $= \dfrac{Sum}{n} = \dfrac{a_1 + a_2 + \ldots + a_n}{n}$

One Geometric Mean
Geometric Mean $= \sqrt[n]{a_1 \cdot a_2 \cdot \ldots \, a_n}$

n Geometric Means Between Two Numbers, a and b	
In a geometric sequence, each term is the geometric mean of the numbers before and after it.	
To find 2 geometric means (m_1, m_2), between a and b, we are looking for a geometric sequence like this: a, m_1, m_2, b. **The common ratio is r.**	
$m_1 = a \cdot r \quad m_2 = a \cdot r^2 \quad b = ar^3 \;\rightarrow\; r = \sqrt[3]{\dfrac{b}{a}}$	
n Geometric Means Between a & b	$r = \sqrt[n+1]{\dfrac{b}{a}}$

Arithmetic Series – Ex. 1

For the given series $\{1, 4, 7, 10, 13, 16, 19, 21\}$
Find:
- Recursive and Explicit equations
- S_5 and S_8
- Arithmetic Mean

Find d	$d = 4 - 1 = 3$
Recursive Eqn.	$a_n = a_{n-1} + d$; $a_1 = $ given $a_n = a_{n-1} + 3$; $a_1 = 1$
Explicit Eqn.	$a_n = a_1 + d(n-1)$ $a_n = 1 + 3(n-1)$
S_n	$S_n = $ Sum of first n terms $S_n = n\left(\frac{a_1 + a_n}{2}\right)$
S_5	$S_5 = 5\left(\frac{a_1 + a_5}{2}\right) = 5\left(\frac{1 + 13}{2}\right) = 35$
S_8	$S_5 = 8\left(\frac{a_1 + a_8}{2}\right) = 8\left(\frac{1 + 21}{2}\right) = 88$
Mean	Average $= \left(\frac{a_1 + a_8}{2}\right) = \left(\frac{1 + 21}{2}\right) = 11$

Arithmetic Series – Ex. 2

For the given series $\{1, 4, 7, 10, 13, \ldots\}$
Find:
- S_{20} = Sum of first 20 terms.
- Arithmetic Mean of the first 20 terms.

Find d	$d = 4 - 1 = 3$	
Explicit Eqn.	$a_n = a_1 + d(n-1)$ $a_n = 1 + 3(n-1)$	
a_{20}	$a_{20} = 1 + 3(20-1) = 58$	
S_n	S_n = Sum of first n terms $S_n = n\left(\dfrac{a_1 + a_n}{2}\right)$	
S_{20}	$S_{20} = 20\left(\dfrac{a_1 + a_{20}}{2}\right)$ $S_{20} = 20\left(\dfrac{1 + 58}{2}\right) = 590$	
Mean	Avg. $= \left(\dfrac{a_1 + a_{20}}{2}\right) = \left(\dfrac{1 + 58}{2}\right) = 29.5$	

Geometric Series – Ex. 3

For the given series $\{\,20, 10, 5, \ldots\,\}$
Find:
- Recursive and Explicit equations
- S_4
- Geometric Mean of the first three terms.

Find r	$r = \dfrac{10}{20} = \dfrac{1}{2}$
Recursive Eqn.	$a_n = a_{n-1}(r)^{n-1}\;;\quad a_1 = given$ $a_n = a_{n-1}\left(\dfrac{1}{2}\right)^{n-1}\;;\quad a_1 = 20$
Explicit Eqn.	$a_n = a_1(r)^{n-1}$ $a_n = 20\left(\dfrac{1}{2}\right)^{n-1}$
S_n	$S_n = $ Sum of first n terms $S_n = a_1\left(\dfrac{1-r^n}{1-r}\right)$
S_4	$S_4 = 20\left(\dfrac{1-r^4}{1-r}\right) = 20(1.875) = 37.5$
G. Mean	$\sqrt[3]{20 \cdot 10 \cdot 5} = \sqrt[3]{1000} = 1000^{\frac{1}{3}} = 10$

Geometric Series – Ex. 4

For the given series $\{20, -10, 5, \ldots\}$
Find:
- Recursive and Explicit equations
- S_4 and S_∞

Find r	$r = \dfrac{-10}{20} = -\dfrac{1}{2}$	Note: $\|r\| < 1$
Recursive Eqn.	$a_n = a_{n-1}(r)^{n-1}$; $\quad a_n = a_{n-1}\left(-\dfrac{1}{2}\right)^{n-1}$;	$a_1 = given$ $a_1 = 20$
Explicit Eqn.	$a_n = a_1(r)^{n-1}$ $a_n = 20\left(-\dfrac{1}{2}\right)^{n-1}$	
S_n	$S_n = a_1\left(\dfrac{1-r^n}{1-r}\right)$	
S_4	$S_4 = 20\left(\dfrac{1-\left(-\frac{1}{2}\right)^4}{1-\left(-\frac{1}{2}\right)}\right) = 20(.625) = 12.5$	
S_∞	$S_\infty = \left(\dfrac{a_1}{1-r}\right) = \left(\dfrac{20}{1-\left(-\frac{1}{2}\right)}\right) = 13.\overline{3}$	

Three Geometric Means – Ex. 5

Find the three geometric means between 10 and 160.

In other words, we are looking for three geometric means (m_1, m_2, m_3) within this geometric series:
$$10, m_1, m_2, m_3, 160$$

Geometric Series	$m_1 = 10 \cdot r$
	$m_2 = 10 \cdot r^2$
	$m_3 = 10 \cdot r^3$
	$160 = 10 \cdot r^4$

Find r	$160 = 10 \cdot r^4$
	$r^4 = \dfrac{160}{10}$
	$r = \left(\dfrac{160}{10}\right)^{\frac{1}{4}} = (16)^{\frac{1}{4}} = 2$

Find the geometric means	$m_1 = 10 \cdot r = 10(2) = 20$
	$m_2 = 10 \cdot r^2 = 10(2)^2 = 40$
	$m_3 = 10 \cdot r^3 = 10(2)^3 = 80$

Matrices – Basic Operations

Matrices – Basic Operations

An $m \times n$ (read "m by n") matrix has m rows and n columns. It is a rectangular array.

$$\begin{bmatrix} a_{11} & a_{12} & a_{13} & \cdots & \cdots & a_{1n} \\ a_{21} & a_{22} & a_{23} & \cdots & \cdots & a_{2n} \\ a_{31} & a_{32} & a_{33} & \cdots & \cdots & a_{3n} \\ \cdot & \cdot & \cdot & \cdots & \cdots & \cdots \\ a_{m1} & a_{m2} & a_{m3} & \cdots & \cdots & a_{mn} \end{bmatrix} \Big\} m \text{ rows}$$

n columns

The **Order** is $m \times n$.

Each entry a_{ij} of the matrix is a real number.
The above matrix has n rows (horizontal lines)
And m columns (vertical lines)

Equality	Two matrices are equal if all of their corresponding entries are equal.
	Given: $A = [a_{ij}]$ and $B = [b_{ij}]$
	$A = B \Leftrightarrow a_{ij} = b_{ij}$

Matrices – Basic Operations
$A = [a_{ij}]$ and $B = [b_{ij}]$

Matrix Addition	$A + B = [a_{ij} + b_{ij}]$ Add corresponding entries.
Scalar Multiplication	$cA = c[a_{ij}] = [c \cdot a_{ij}]$ Multiply every entry by "c"

Matrix Multiplication

$$\begin{bmatrix} 1 & 2 & 3 \\ 4 & 5 & 6 \end{bmatrix} \cdot \begin{bmatrix} 2 & 8 \\ 4 & 10 \\ 6 & 12 \end{bmatrix} = 2 \times 2 \text{ Matrix}$$

$2 \times \boxed{3} \cdot \boxed{3} \times 2 \quad \rightarrow \quad 2 \times 2 \text{ Matrix}$

$$= \begin{vmatrix} \begin{array}{cc} 1 \cdot 2 + & 1 \cdot 8 + \\ 2 \cdot 4 + & 2 \cdot 10 + \\ 3 \cdot 6 & 3 \cdot 12 \end{array} \\ \begin{array}{cc} 4 \cdot 2 + & 4 \cdot 8 + \\ 5 \cdot 4 + & 5 \cdot 10 + \\ 6 \cdot 6 & 6 \cdot 12 \end{array} \end{vmatrix} = \begin{vmatrix} \begin{array}{cc} 2 + & 8 + \\ 8 + & 20 + \\ 18 & 36 \end{array} \\ \begin{array}{cc} 8 + & 32 + \\ 20 + & 50 + \\ 36 & 72 \end{array} \end{vmatrix}$$

$$= \begin{bmatrix} 28 & 64 \\ 64 & 154 \end{bmatrix}$$

Matrices – Properties Addition and Scalar Multiplication	
Let A, B, C be $m \times n$ (m rows, n columns) matrices. Let c and d be scalars.	
1.	$A + B = B + A$
2.	$A + (B + C) = (A + B) + C$
3.	$(cd)A = c(dA)$
4.	$1A = A$
5.	$c(A + B) = cA + cB$
6.	$(c + d)A = cA + dA$

Matrix Multiplication – Properties	
1.	$A(BC) = (AB)C$
2.	$A(B + C) = AB + AC$
3.	$(A + B)C = AC + BC$
4.	$c(AB) = (cA)B = A(cB)$

Identity Matrix

The **Identity Matrix** for a $n \times n$ square matrix is:

$$I = \begin{bmatrix} 1 & 0 & 0 \\ 0 & 1 & 0 \\ 0 & 0 & 1 \end{bmatrix}$$

It has 1's on the diagonal. 0's in all other entries.
$$IA = A \quad \text{and} \quad AI = A$$

Identity Matrix Examples

Given: $A = \begin{bmatrix} 1 & 2 \\ 3 & 4 \end{bmatrix} \quad I = \begin{bmatrix} 1 & 0 \\ 0 & 1 \end{bmatrix}$

Find: AI and IA

AI:
$$AI = \begin{bmatrix} 1 & 2 \\ 3 & 4 \end{bmatrix} \cdot \begin{bmatrix} 1 & 0 \\ 0 & 1 \end{bmatrix}$$
$$AI = \begin{bmatrix} (1+0) & (0+2) \\ (3+0) & (0+4) \end{bmatrix} = \begin{bmatrix} 1 & 2 \\ 3 & 4 \end{bmatrix} = A$$

IA:
$$IA = \begin{bmatrix} 1 & 0 \\ 0 & 1 \end{bmatrix} \cdot \begin{bmatrix} 1 & 2 \\ 3 & 4 \end{bmatrix}$$
$$IA = \begin{bmatrix} (1+0) & (2+0) \\ (0+3) & (0+4) \end{bmatrix} = \begin{bmatrix} 1 & 2 \\ 3 & 4 \end{bmatrix} = A$$

The Inverse of a Square Matrix, 2 × 2

A^{-1} = The inverse of matrix A

$A \cdot A^{-1} = I$

If $A = \begin{bmatrix} a & b \\ c & d \end{bmatrix}$ Then $A^{-1} = \frac{1}{ad - bc} \begin{bmatrix} d & -b \\ -c & a \end{bmatrix}$

Note: A is invertible only if $ad - bc \neq 0$

Identity Matrix Example

Given: $A = \begin{bmatrix} 1 & 2 \\ 3 & 4 \end{bmatrix}$ Find: A^{-1} and $A \cdot A^{-1}$

A^{-1}	$A^{-1} = \frac{1}{4-6} \begin{bmatrix} 4 & -2 \\ -3 & 1 \end{bmatrix} = -\frac{1}{2} \begin{bmatrix} 4 & -2 \\ -3 & 1 \end{bmatrix}$ $A^{-1} = \begin{bmatrix} -2 & 1 \\ \frac{3}{2} & -\frac{1}{2} \end{bmatrix}$
$A \cdot A^{-1}$	$A \cdot A^{-1} = \begin{bmatrix} 1 & 2 \\ 3 & 4 \end{bmatrix} \cdot \begin{bmatrix} -2 & 1 \\ \frac{3}{2} & -\frac{1}{2} \end{bmatrix}$ $A \cdot A^{-1} = \begin{bmatrix} (-2+3) & (1-1) \\ (-6+6) & (3-2) \end{bmatrix}$ $A \cdot A^{-1} = \begin{bmatrix} 1 & 0 \\ 0 & 1 \end{bmatrix} = I$

Matrices – Basic Operations – Ex. 1

Do the indicated matrix operations, given:

$$A = \begin{bmatrix} 1 & 2 \\ 3 & 4 \end{bmatrix} \qquad B = \begin{bmatrix} -1 & 3 \\ -2 & 4 \end{bmatrix} \qquad C = \begin{bmatrix} 0 & 1 & 1 \\ 2 & 0 & 3 \end{bmatrix}$$

$A + B$	$= \begin{bmatrix} (1-1) & (2+3) \\ (3-2) & (4+4) \end{bmatrix} = \begin{bmatrix} 0 & 5 \\ 1 & 16 \end{bmatrix}$
$A + C$	Not possible. Must be same size.
AB	$2 \times \boxed{2} \quad \boxed{2} \times 2 \;\rightarrow\; 2 \times 2$ $AB = \begin{bmatrix} 1 & 2 \\ 3 & 4 \end{bmatrix} \cdot \begin{bmatrix} -1 & 3 \\ -2 & 4 \end{bmatrix}$ $AB = \begin{bmatrix} (-1-4) & (3+8) \\ (-3-8) & (9+16) \end{bmatrix}$ $AB = \begin{bmatrix} -5 & 11 \\ -11 & 25 \end{bmatrix}$
AC	$2 \times \boxed{2} \quad \boxed{2} \times 3 \;\rightarrow\; 2 \times 3$ $AC = \begin{bmatrix} 1 & 2 \\ 3 & 4 \end{bmatrix} \cdot \begin{bmatrix} 0 & 1 & 1 \\ 2 & 0 & 3 \end{bmatrix}$ $AC = \begin{bmatrix} (0+4) & (1+0) & (1+6) \\ (0+8) & (3+0) & (3+12) \end{bmatrix}$ $AC = \begin{bmatrix} 4 & 1 & 7 \\ 8 & 3 & 15 \end{bmatrix}$

Matrices – Systems of Eqns.

Systems of Equations in Matrix Form	
Systems of equations can be represented as matrices. • Coefficients are stored in the coefficient matrix. • Variables are stored in the variable matrix. • Answers are stored in the answer matrix.	
Consider a system of 2 equations with 2 unknowns.	$2x + 3y = 8$ $x + 2y = 5$
Store coefficients in a coefficient matrix	$C = \begin{bmatrix} 2 & 3 \\ 1 & 2 \end{bmatrix}$
Store variables in a variable matrix.	$X = \begin{bmatrix} x \\ y \end{bmatrix}$
Store answers in an answer matrix.	$B = \begin{bmatrix} 8 \\ 5 \end{bmatrix}$
Matrix Equation	$CX = B$

$$CX = \begin{bmatrix} 2 & 3 \\ 1 & 2 \end{bmatrix} \cdot \begin{bmatrix} x \\ y \end{bmatrix} = \begin{bmatrix} 8 \\ 5 \end{bmatrix}$$

$$2 \times ②\ \ ②\times 1 \Leftrightarrow 2 \times 1$$

Augmented Matrix
Augmented matrices include the coefficients and answers (no variables).

Consider a system of 2 equations with 2 unknowns.	$2x + 3y = 8$ $x + 2y = 5$
Augmented Matrix	$\begin{bmatrix} 2 & 3 & 8 \\ 1 & 2 & 5 \end{bmatrix}$
Optional: A dotted line may be added	$\begin{bmatrix} 2 & 3 & \vdots & 8 \\ 1 & 2 & \vdots & 5 \end{bmatrix}$

Row Operations on Matrices

Matrices can be simplified with row operations. (Add, Subtract, Multiply, Divide)   The goal is to get: 1's on the diagonal and 0's most everywhere else. Make little notes about what you do, along the way!

Original Matrix	$\begin{bmatrix} 2 & 3 & 8 \\ 1 & 2 & 5 \end{bmatrix}$
Switch Rows	$\begin{bmatrix} 2 & 3 & 8 \\ 1 & 2 & 5 \end{bmatrix} \xrightarrow{r_1 \leftrightarrow r_2} \begin{bmatrix} 1 & 2 & 5 \\ 2 & 3 & 8 \end{bmatrix}$
Multiply & Subt.	$\begin{bmatrix} 1 & 2 & 5 \\ 2 & 3 & 8 \end{bmatrix} \xrightarrow{r_2 - 2r_1} \begin{bmatrix} 1 & 2 & 5 \\ 0 & -1 & -2 \end{bmatrix}$
Multiply & Add.	$\begin{bmatrix} 1 & 2 & 5 \\ 0 & -1 & -2 \end{bmatrix} \xrightarrow{r_1 + 2r_2} \begin{bmatrix} 1 & 0 & 1 \\ 0 & -1 & -2 \end{bmatrix}$
Multiply	$\begin{bmatrix} 1 & 2 & 5 \\ 0 & -1 & -2 \end{bmatrix} \xrightarrow{(-1)r_2} \begin{bmatrix} 1 & 0 & 1 \\ 0 & 1 & 2 \end{bmatrix}$
Solution	$x = 1$, $y = 2$

In General	$\begin{bmatrix} 1 & 0 & a \\ 0 & 1 & b \end{bmatrix}$	$x = a$, $y = b$

Row Echelon Form -- ref	
1.	All rows, consisting entirely of zeros, occur at the bottom of the matrix.
2.	For each, non-all-zero row, the first non-zero entry is 1 (called the leading 1).
3.	For two successive non-all-zero rows, the leading 1 in the higher row is to the left of the leading 1 in th lower row.

Reduced Row Echelon Form – rref	
4.	Every column that has a leading 1 has 0's in every position above and below its leading 1.

Row Echelon Form -- Examples	
$\begin{bmatrix} 1 & 2 & -1 & 4 \\ 0 & 1 & 0 & 3 \\ 0 & 0 & 1 & -2 \end{bmatrix}$	$\begin{bmatrix} 0 & 1 & 0 & 5 \\ 0 & 0 & 1 & 3 \\ 0 & 0 & 0 & 0 \end{bmatrix}$ Also in reduced row echelon form.
$\begin{bmatrix} 1 & -5 & 2 & -1 & 3 \\ 0 & 0 & 1 & 3 & -2 \\ 0 & 0 & 0 & 1 & 4 \\ 0 & 0 & 0 & 0 & 1 \end{bmatrix}$	$\begin{bmatrix} 1 & 0 & 0 & -1 \\ 0 & 1 & 0 & 2 \\ 0 & 0 & 1 & 3 \\ 0 & 0 & 0 & 0 \end{bmatrix}$ Also in reduced row echelon form.

Gaussian Elimination with Back-Substitution Example	
Solve the system	$y + z - 2w = -3$ $x + 2y - z = 2$ $2x + 4y + z - 3w = -2$ $x - 4y - 7z - w = -19$
Augmented Matrix	$\begin{bmatrix} 0 & 1 & 1 & -2 & -3 \\ 1 & 2 & -1 & 0 & 2 \\ 2 & 4 & 1 & -3 & -2 \\ 1 & -4 & -7 & -1 & -19 \end{bmatrix}$
$r_1 \leftrightarrow r_2$	$\begin{bmatrix} 1 & 2 & -1 & 0 & 2 \\ 0 & 1 & 1 & -2 & -3 \\ 2 & 4 & 1 & -3 & -2 \\ 1 & -4 & -7 & -1 & -19 \end{bmatrix}$
$r_3 - 2r_1$ $r_4 - r_1$	$\begin{bmatrix} 1 & 2 & -1 & 0 & 2 \\ 0 & 1 & 1 & -2 & -3 \\ 0 & 0 & 3 & -3 & -6 \\ 0 & -6 & -6 & -1 & -21 \end{bmatrix}$
$r_4 + 6r_2$	$\begin{bmatrix} 1 & 2 & -1 & 0 & 2 \\ 0 & 1 & 1 & -2 & -3 \\ 0 & 0 & 3 & -3 & -6 \\ 0 & 0 & 0 & -13 & -39 \end{bmatrix}$
Continued …	

	Gaussian Elimination with Back-Substitution Example (Continued)
Solve the system	$y + z - 2w = -3$ $x + 2y - z = 2$ $2x + 4y + z - 3w = -2$ $x - 4y - 7z - w = -19$
Previously Found	$\begin{bmatrix} 1 & 2 & -1 & 0 & 2 \\ 0 & 1 & 1 & -2 & -3 \\ 0 & 0 & 3 & -3 & -6 \\ 0 & 0 & 0 & -13 & -39 \end{bmatrix}$
$\frac{1}{3} r_3$	$\begin{bmatrix} 1 & 2 & -1 & 0 & 2 \\ 0 & 1 & 1 & -2 & -3 \\ 0 & 0 & 1 & -1 & -2 \\ 0 & 0 & 0 & -13 & -39 \end{bmatrix}$
$-\frac{1}{13} r_4$	$\begin{bmatrix} 1 & 2 & -1 & 0 & 2 \\ 0 & 1 & 1 & -2 & -3 \\ 0 & 0 & 1 & -1 & -2 \\ 0 & 0 & 0 & 1 & 3 \end{bmatrix}$
Result in Row Echelon Form. Corresponding system is:	$x + 2y - z = 2$ $y + z - 2w = -3$ $z - w = -2$ $w = 3$
Solution	$(x, y, z, w) = (-1, 2, 1, 3)$

Matrix Calculations on Desmos

A matrix calculator is available online at
www.desmos.com/matrix

Desmos can be used to check your answers!
Desmos solution to previous problem is shown below.

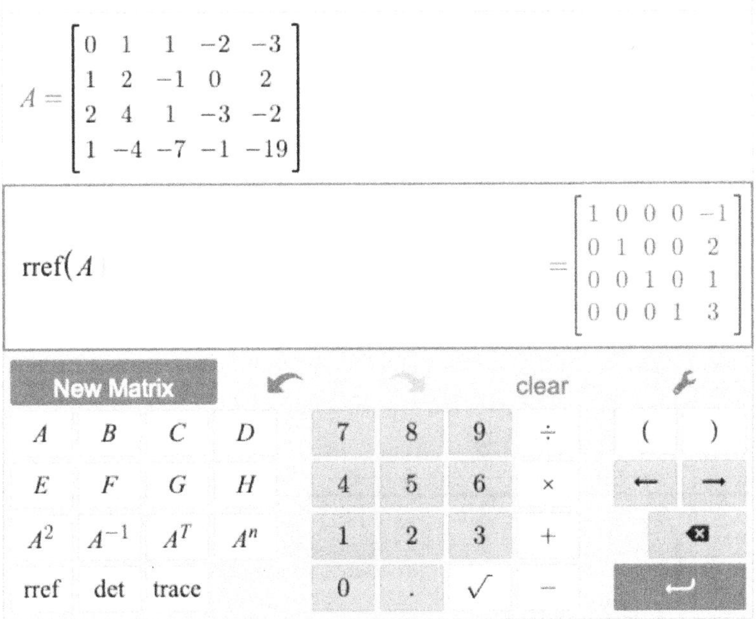

$$A = \begin{bmatrix} 0 & 1 & 1 & -2 & -3 \\ 1 & 2 & -1 & 0 & 2 \\ 2 & 4 & 1 & -3 & -2 \\ 1 & -4 & -7 & -1 & -19 \end{bmatrix}$$

$$\text{rref}(A) = \begin{bmatrix} 1 & 0 & 0 & 0 & -1 \\ 0 & 1 & 0 & 0 & 2 \\ 0 & 0 & 1 & 0 & 1 \\ 0 & 0 & 0 & 1 & 3 \end{bmatrix}$$

Matrix Calculations on TI Calculator

For a Ti-84 Plus CE, the commands are:
- 2nd matrix, EDIT To input a matrix.
- 2nd matrix, MATH, scroll down to rref(
- To specify matrix [A] into rref(, do this:
    2nd matrix, NAMES , then close parens.

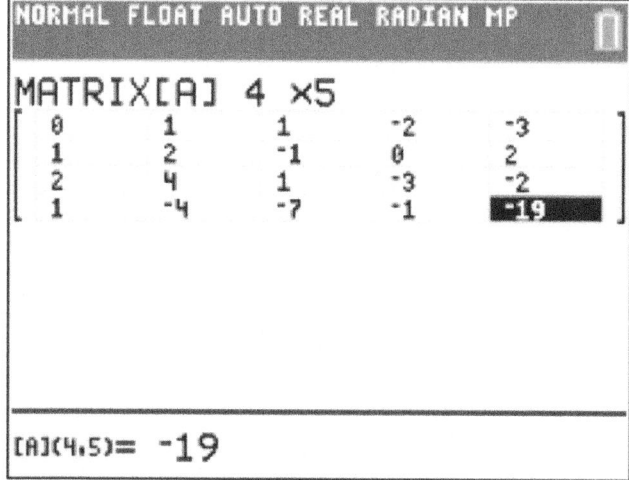

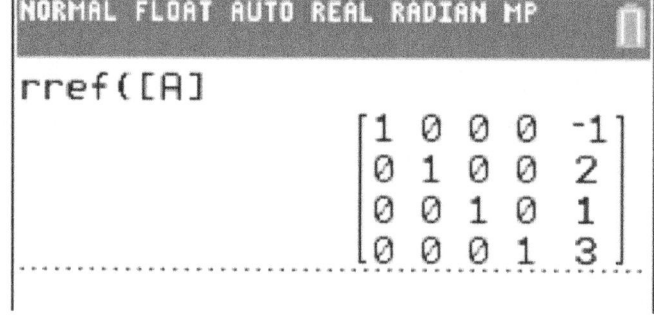

Matrices – More Examples

Matrices – More Examples

The following examples are included in this section:

#	Example
1	Create a simple 2x2 matrix then solve it • With a TI calculator and • With Desmos/matrix
2	Simple matrix with two types of errors: • One row is a multiple of another • One row is based on an incorrect equation.
3	Coin example with 3 types of coins.
4	Coin example with 4 types of coins.
5	Use a matrix to solve a system of equations.
6	Use a matrix to do linear regression.
7	Use a matrix to do quadratic regression.
8	Use a matrix to do cubic regression.

Matrices – Simple Example – 1a
Create a system of equations with 2 equations and 2 unknowns. Then, solve it.

Define 2 unknowns	x and y
	Let $x = 2$ and $y = 5$
Write some equations	$x + y = 7$ $x - y = -3$ $y - 2x = 1$ $y = 5$ ...
With 2 unknowns, only need 2 equations are needed. Pick 2.	$x - y = -3$ $y - 2x = 1$
Put equations in matrix form. Create a 2x3 augmented matrix. 2 rows, 3 cols.	$x - y = -3$ $-2x + y = 1$ $\begin{array}{ccc} x & y & = n \end{array}$ $\begin{bmatrix} 1 & -1 & \vdots & -3 \\ -2 & 1 & \vdots & 1 \end{bmatrix}$

Matrices – Simple Example -- 1b
Create a system of equations with 2 equations and 2 unknowns. Then, solve it.

2x3 Augmented Matrix Created previously	$\begin{bmatrix} 1 & -1 & -3 \\ -2 & 1 & 1 \end{bmatrix}$

Solve with TI-84 Calculator	
Input 2x3 Matrix	[2nd] matrix, EDIT MATRIX[A] 2 ×3 $\begin{bmatrix} 1 & -1 & -3 \\ -2 & 1 & 1 \end{bmatrix}$
Solve Matrix	[2nd] matrix, MATH, rref rref([2nd] matrix, NAMES, [A], [enter] rref([A] rref([A]) $\begin{bmatrix} 1 & 0 & 2 \\ 0 & 1 & 5 \end{bmatrix}$
Answer In last column	$(x, y) = (2, 5)$ Correct!

Matrices – Simple Example – 1c
Create a system of equations with 2 equations and 2 unknowns. Then, solve it.

2x3 Augmented Matrix Created previously	$\begin{bmatrix} 1 & -1 & -3 \\ -2 & 1 & 1 \end{bmatrix}$

Solve with Desmos: www.Desmos.com/matrix	
Input New Matrix 2 Rows 3 Columns	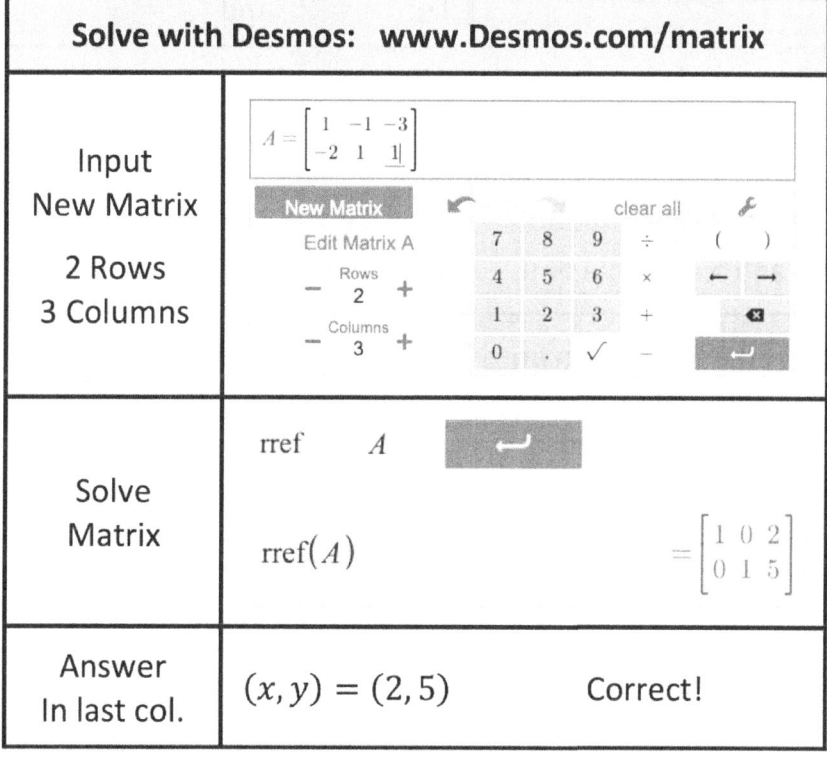
Solve Matrix	rref A ⏎ rref(A) $= \begin{bmatrix} 1 & 0 & 2 \\ 0 & 1 & 5 \end{bmatrix}$
Answer In last col.	$(x, y) = (2, 5)$ Correct!

Matrices – Simple Example Errors – Ex. 2a	
What would happen if the system of equations, in the previous example, had a mistake in it? In this example, two kinds of mistakes will be reviewed.	
Previously Defined 2 unknowns	x and y
	Let $x = 2$ and $y = 5$

Mistake #1: One equation is a multiple.	
Write one equation and a multiple of it.	$y - 2x = 1$ $2y - 4x = 2$ Multiple
Put equations in an augmented matrix	$\begin{bmatrix} 1 & -2 & 1 \\ 2 & -4 & 2 \end{bmatrix}$

Mistake #2: One equation has a mistake in it.	
Write one good and one bad equation.	$y - 2x = 1$ Good $y - 2x = 7$ Bad
Put equations in an augmented matrix	$\begin{bmatrix} 1 & -2 & 1 \\ 1 & -2 & 7 \end{bmatrix}$

Matrices – Simple Example Errors – Ex. 2b

Mistake #1: One equation is a multiple.

| Previously created error. 2nd row is a multiple of 1st row. | $\begin{bmatrix} 1 & -2 & 1 \\ 2 & -4 & 2 \end{bmatrix}$ |

Try to Solve with TI-84 Calculator

[2nd] matrix, MATH, rref   [2nd] matrix, NAMES, [A]

$$\text{rref([A])}$$
$$\begin{bmatrix} 1 & \text{-}2 & 1 \\ 0 & 0 & 0 \end{bmatrix}$$

Try to Solve with Desmos Matrix Calculator

$\text{rref}(A| \qquad\qquad = \begin{bmatrix} 1 & -2 & 1 \\ 0 & 0 & 0 \end{bmatrix}$

The matrix cannot be solved. When solving for n variables, n distinct equations are needed.

Matrices – Simple Example Errors – Ex. 2c

Mistake #2: One equation has a mistake in it.

Previously created error. Matrix with mistake.	$\begin{bmatrix} 1 & -2 & 1 \\ 1 & -2 & 7 \end{bmatrix}$

Try to Solve with TI-84 Calculator

[2nd] matrix, MATH, rref   [2nd] matrix, NAMES, [A]

rref([A])

$\begin{bmatrix} 1 & -2 & 0 \\ 0 & 0 & 1 \end{bmatrix}$

Try to Solve with Desmos Matrix Calculator

$\text{rref}(A)$ $= \begin{bmatrix} 1 & -2 & 0 \\ 0 & 0 & 1 \end{bmatrix}$

The matrix cannot be solved. It says:
$x - 2y = 0$ and $0x + 0y = 1$
The last row implies that $0 = 1$.
This means the sytem is **inconsistent**.

	Matrices – Coin Example – Ex.3
	Create a system of 3 equations and 3 unknowns. Then, solve it. Use pennies, nickels, and dimes.

Define 3 unknowns	$p = \# \, pennies = 4$ $n = \# \, nickels = 3$ $d = \# \, dimes = 1$
Write 3 equations	$p + n + d = 8$ $p(1) + n(5) + d(10) = 29$ $p(1) + n(5) = 19$
Note: There are 8 coins with a total value of 29 cents.	
Put equations in matrix form. Augmented matrix. 3 rows, 4 cols.	$\begin{array}{cccc} p & n & d & = \# \end{array}$ $\begin{bmatrix} 1 & 1 & 1 & \vdots & 8 \\ 1 & 5 & 10 & \vdots & 29 \\ 1 & 5 & 0 & \vdots & 19 \end{bmatrix}$
Solve with TI-84 Calculator	[2nd] matrix, EDIT, MATH, rref rref([A]) $\begin{bmatrix} 1 & 0 & 0 & 4 \\ 0 & 1 & 0 & 3 \\ 0 & 0 & 1 & 1 \end{bmatrix}$
Answer	$(p, n, d) = (4, 3, 1)$ Correct!

Matrices – Another Coin Example – Ex. 4

Suppose you have 10 coins in your pocket.
The coins are pennies, nickels, dimes, and quarters.
The total value of your coins is 78 cents.
The value of the nickels and dimes is 50 cents.
Two of the coins are nickels.
Find the number of each type of coin.

Write 4 equations	$p + n + d + q = 10$ $p(1) + n(5) + d(10) + q(25) = 78$ $n(5) + d(10) = 50$ $n = 2$
Create Augmented Matrix (4 x 5)	$\begin{array}{cccc} p & n & d & q = \# \end{array}$ $\begin{bmatrix} 1 & 1 & 1 & 1 & \mid 10 \\ 1 & 5 & 10 & 25 & \mid 78 \\ 0 & 5 & 10 & 0 & \mid 50 \\ 0 & 1 & 0 & 0 & \mid 2 \end{bmatrix}$
Solve with Desmos Matrix	$\text{rref}(A) = \begin{bmatrix} 1 & 0 & 0 & 0 & 3 \\ 0 & 1 & 0 & 0 & 2 \\ 0 & 0 & 1 & 0 & 4 \\ 0 & 0 & 0 & 1 & 1 \end{bmatrix}$
Answer	$(p, n, d, q) = (3, 2, 4, 1)$

Using Matrices to Solve Systems of Linear Equations — Ex. 5

Given this system: $\begin{cases} x - 2y = 5 \\ 4x + 3y = 9 \end{cases}$

Use Desmos/Matrix or a calculator to solve it.

Desmos/Matrix Solution
(www.desmos.com/matrix)

$A = \begin{bmatrix} 1 & -2 & 5 \\ 4 & 3 & 9 \end{bmatrix}$

$\text{rref}(A) = \begin{bmatrix} 1 & 0 & 3 \\ 0 & 1 & -1 \end{bmatrix}$

Answer in last column.

$(x, y) = (3, -1)$

Using Matrices to do Linear Regression – Ex. 6

Given two points: $(3, 2)$, $(7, 10)$
Find: Linear equation that goes through them.

Use given points to setup an augmented matrix.	$y = ax + b$
	$2 = a(3) + b$ $10 = a(7) + b$
	$\begin{bmatrix} 3 & 1 & 2 \\ 7 & 1 & 10 \end{bmatrix}$
Use Desmos /Matrix or Calculator to find a & b	$A = \begin{bmatrix} 3 & 1 & 2 \\ 7 & 1 & 10 \end{bmatrix}$
	$\text{rref}(A) \qquad\qquad = \begin{bmatrix} 1 & 0 & 2 \\ 0 & 1 & -4 \end{bmatrix}$
	$(a, b) = (2, -4)$
Linear Eqn.	$y = ax + b$ $y = 2x - 4$
Graph	

Using Matrices to do Quadratic Regression – Ex. 7	
Given three points: $(-2, 3)$, $(1, 6)$, $(5, 38)$ Find: Quadratic equation that goes through them.	
Use given points to setup an augmented matrix.	$y = ax^2 + bx + c$
	$3 = a(-2)^2 + b(-2) + c$ $6 = a(1)^2 + b(1) + c$ $38 = a(5)^2 + b(5) + c$
	$\begin{bmatrix} 4 & -2 & 1 & 3 \\ 1 & 1 & 1 & 6 \\ 25 & 5 & 1 & 38 \end{bmatrix}$
Use Desmos /Matrix or Calculator to find a, b & c	$A = \begin{bmatrix} 4 & -2 & 1 & 3 \\ 1 & 1 & 1 & 6 \\ 25 & 5 & 1 & 38 \end{bmatrix}$
	$\text{rref}(A) = \begin{bmatrix} 1 & 0 & 0 & 1 \\ 0 & 1 & 0 & 2 \\ 0 & 0 & 1 & 3 \end{bmatrix}$
	$(a, b, c) = (1, 2, 3)$
Quad. Eqn.	$y = ax^2 + bx + c$ $y = x^2 + 2x + 3$

Using Matrices to do Cubic Regression – Ex. 8

Given points: $(1,-2)$, $(0,-4)$, $(-1,-10)$, $(5,36)$
Find: Cubic equation that goes through them.

	$y = ax^3 + bx^2 + cx + d$
Use given points to setup an augmented matrix.	$-2 = a(1)^3 + b(1)^2 + c(1) + d$ $-4 = a(0)^3 + b(0)^2 + c(0) + d$ $-10 = a(-1)^3 + b(-1)^2 + c(-1) + d$ $36 = a(5)^3 + b(5)^2 + c(5) + d$
	$\begin{bmatrix} 1 & 1 & 1 & 1 & -2 \\ 0 & 0 & 0 & 1 & -4 \\ -1 & 1 & -1 & 1 & -10 \\ 75 & 25 & 5 & 1 & 36 \end{bmatrix}$
Use Desmos /Matrix or Calculator to find constants	$A = \begin{bmatrix} 1 & 1 & 1 & 1 & -2 \\ 0 & 0 & 0 & 1 & -4 \\ -1 & 1 & -1 & 1 & -10 \\ 75 & 25 & 5 & 1 & 36 \end{bmatrix}$ $\text{rref}(A) = \begin{bmatrix} 1 & 0 & 0 & 0 & 1 \\ 0 & 1 & 0 & 0 & -2 \\ 0 & 0 & 1 & 0 & 3 \\ 0 & 0 & 0 & 1 & -4 \end{bmatrix}$ $(a,b,c,d) = (1,-2,\ 3,-4)$
Cubic Eqn.	$y = x^3 - 2x^2 + 3x - 4$

Statistics

Basic Statistical Data	
Mean (Average)	$\bar{x} = \dfrac{\sum_{i=1}^{n} x_i}{n}$
Median	Middle term or average of middle terms
Mode	Term(s) occurring most often

Example	
\multicolumn{2}{c}{Given: $\{0, 1, 7, 9, 4, 5, 2, 7\}$}
Put data in order	$\{0, 1, 2, 4, 5, 7, 7, 9\}$, $n = 8$
Mean (Average)	$\bar{x} = \dfrac{0+1+2+4+5+7+7+9}{8}$ $\bar{x} = \dfrac{35}{8} = 4.375$
Median	$0, 1, 2, 4 \mid 5, 7, 7, 9$ $\dfrac{4+5}{2} = 4.5$
Mode	7 occurs most often

(Example row 2: Find: Mean, Median, Mode)

Presenting Statical Data
Frequency Distribution and Histograms

Frequency Distribution	A table, showing values and how often that value occurs.
Histogram	A graphical representation of a frequency distribution.

Example

Value	Count
1	1
5	2
10	1
18	5
21	4
31	6
40	10
52	8
66	10
71	6
75	4
88	3
92	1
96	2
99	1

The horizontal axis shows the value. The vertical axis shows the frequency (or count).

For a better histogram, set ranges of values on the x-axis. See next page.

Presenting Statical Data
Frequency Distribution and Histograms

Value	Count
1	1
5	2
10	1
18	5
21	4
31	6
40	10
52	8
66	10
71	6
75	4
88	3
92	1
96	2
99	1

How to create a histogram on a TI-84 Plus CE graphing calculator.

- [stat], EDIT.
- Enter values in L1, freq. in L2
- 2nd [y=] Stat Plot , ENTER "on"
- Select histogram icon.
- Enter Freq. 2nd L2

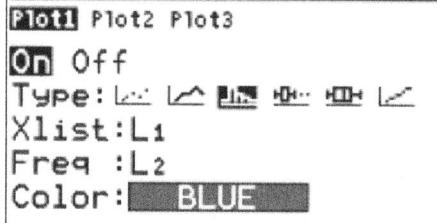

- [window] , xscl=10
- [graph], [trace]

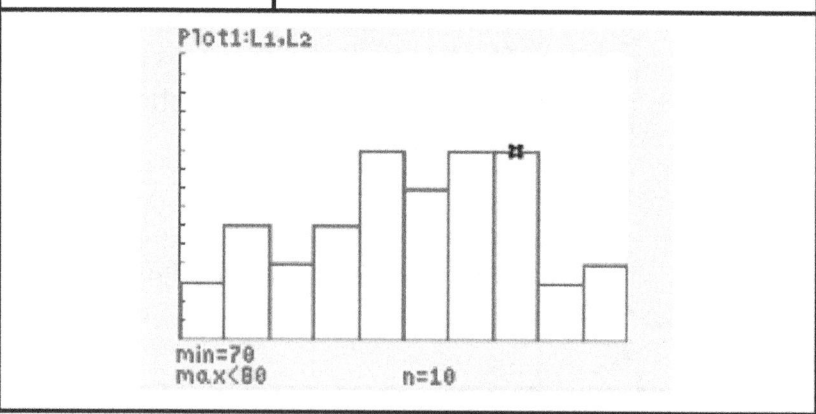

\multicolumn{2}{c}{**Presenting Statical Data**}
\multicolumn{2}{c}{**Stem and Leaf**}
Stem and Leaf Diagram	Another way to present data that includes the data in the display.
Statistics	• **Mean**: Average of all values. • **Median**: Middle value or average of middle values • **Mode**: Value or values occurring most often.

Example

Values
11
18
18
22
22
22
28
33
34
40
45
46

1	1	8	8	
2	2	2	2	2
3	3	4		
4	0	5	6	

- The shape of the stem & leaf diagram has the shape of a sideways histogram.
- The first column is the stem and represents 10's.
- The leaves represent 1's.

Presenting Statical Data
Statistics

Using the following values, we can calculate statistics.

| 1 | 8 | 3 | 2 | 7 | 4 | 5 | 6 | 3 | 3 |

Put the data in order.	1 2 3 3 3 | 4 5 6 7 8
Number	$n = 10$
Min.	Lowest value $= 1$
Max.	Largest value $= 8$
Mode	Value occurring most often $= 3$
Median	Middle value $= \dfrac{3+4}{2} = \dfrac{7}{2} = 3.5$
Mean	$\bar{x} = \dfrac{1 + 2 + 3(3) + 4 + 5 + 6 + 7 + 8}{10}$ $\bar{x} = 4.2$

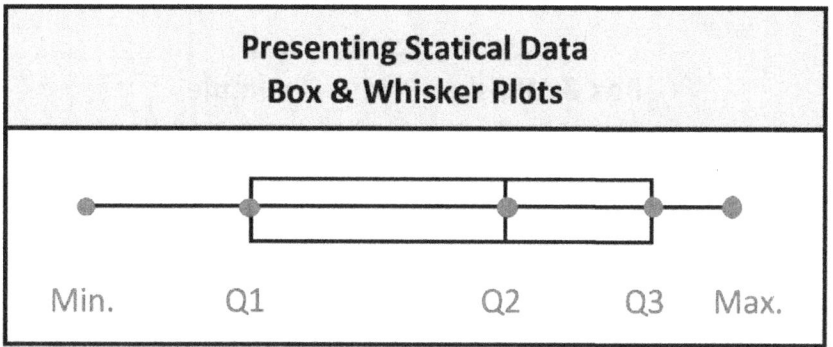

First Quartile	Min. – Q1
Second Quartile	Q1 – Q2
Third Quartile	Q2 – Q3
Fourth Quartile	Q3 – Max.

Q2	Median (middle) of all data.
Q1	Median (middle) of lower half of data.
Q3	Median (middle) of upper half of data.

Presenting Statical Data
Box & Whisker Plots -- Example

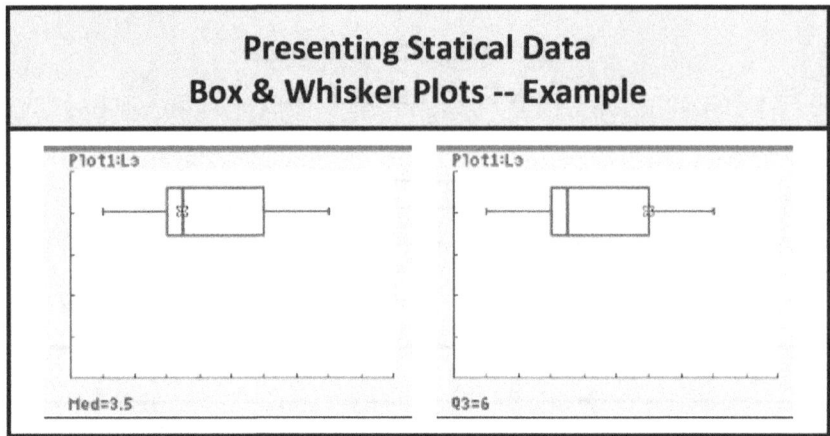

Example	
Put data in order.	1 2 3 3 3 4 5 6 7 8
Number	$n = 10$
Min.	Lowest value $= 1$
Max.	Largest value $= 8$
Median	Middle value $= \dfrac{3+4}{2} = \dfrac{7}{2} = 3.5$
Q1	Middle of lower half $= 3$
Q2	Median $= 3.5$
Q3	Middle of upper half $= 6$

Presenting Statical Data
Statistics on TI-84 Graphing Calculator

Using this data.	1	2	3	3	3	4	5	6	7	8

Put data in a list then specify plot parameters.	[stat], EDIT 2nd] [y=] Stat Plot
Select the box & whisker icon. Select the list. [2nd] L3 in this example.	**Plot1** Plot2 Plot3 ■ Off Type: ⌐ ⌐ ⌐ ⌐ ⌐ ⌐ Xlist: L3 Freq : 1 Color: ■ BLUE
Get statistics. [STAT], CALC, 1-Var Stats	**1-Var Stats** List: L3 FreqList: Calculate

1-Var Stats
$\bar{x}=4.2$
$\Sigma x=42$
$\Sigma x^2=222$
$Sx=2.250925735$
$\sigma x=2.13541565$
$n=10$
$minX=1$
↓$Q_1=3$

1-Var Stats
↑$\Sigma x=42$
$\Sigma x^2=222$
$Sx=2.250925735$
$\sigma x=2.13541565$
$n=10$
$minX=1$
$Q_1=3$
↓$Med=3.5$

1-Var Stats
↑$\Sigma x^2=222$
$Sx=2.250925735$
$\sigma x=2.13541565$
$n=10$
$minX=1$
$Q_1=3$
$Med=3.5$
↓$Q_3=6$

1-Var Stats
↑$Sx=2.250925735$
$\sigma x=2.13541565$
$n=10$
$minX=1$
$Q_1=3$
$Med=3.5$
$Q_3=6$
$maxX=8$
■

Normal Distribution -- Population Parameters	
Mean (Average)	$\mu = \dfrac{\sum_{i=1}^{N} x_i}{N}$
Variance	$\sigma^2 = \dfrac{\sum_{i=1}^{N} (x_i - \bar{x})^2}{N}$
Standard Deviation	σ

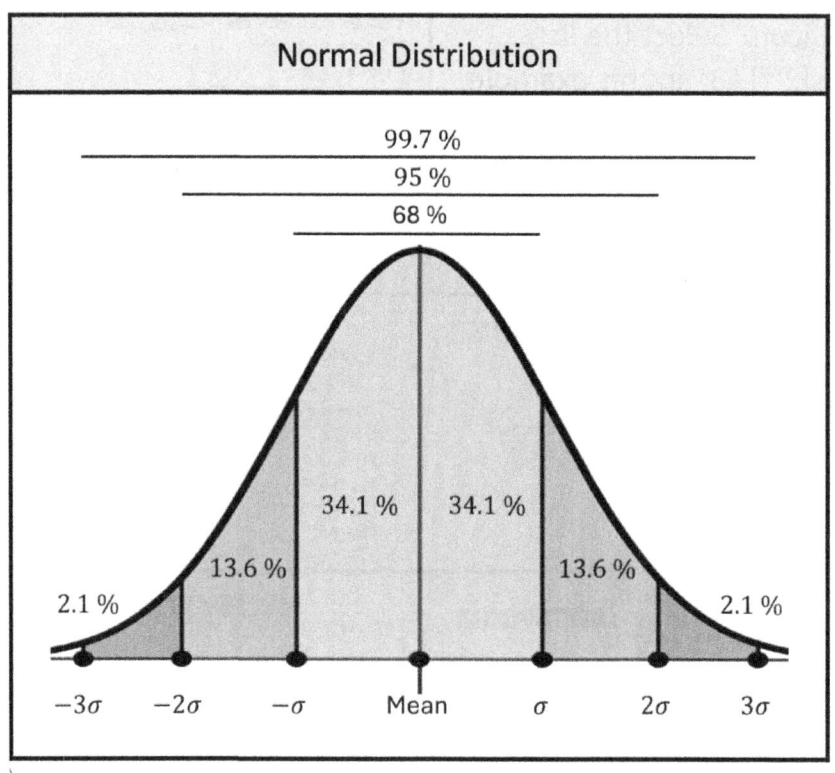

Normal Distribution – Sample Statistics	
	Population statistics may not be available. Calculating statistics for a sample may be easier.
Mean (Average)	$\bar{x} = \dfrac{\sum_{i=1}^{n} x_i}{n}$
Variance	$s^2 = \dfrac{\sum_{i=1}^{n} (x_i - \bar{x})^2}{n}$
Standard Deviation	s
Median	Average of middle term(s).
Mode	Term(s) occurring most often
Z Score	$z_i = \dfrac{x_i - \mu}{\sigma} = \dfrac{x_i - \bar{x}}{s}$

Normal Distribution of Sample
If the sample data has a normal distribution, then you can expect: $$\bar{x} \pm 1s = 68\%$$ $$\bar{x} \pm 2s = 95\%$$ $$\bar{x} \pm 3s = 99.7\%$$

Presenting Statical Data
Box & Whisker Plots – Example 1

Two classes took the same algebra test. The results are shown in the box-and-whisker plots below.

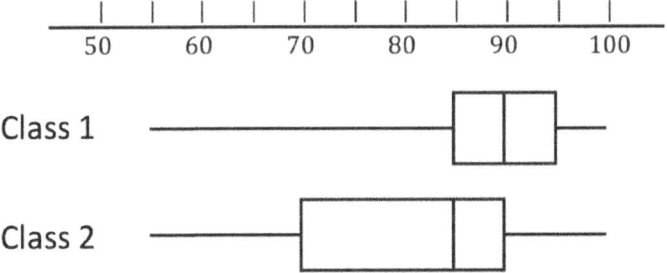

A. Which class has the higher median?
B. Which class has the smaller range?
C. For which class are the scores in the middle-half closer together?
D. Which class has the better set of scores?

A	Class 1 Median: 90 Class 2 Median: 85
B	Class 1 Range: 45 Class 2 Range: 45 Same
C	Box shows the scores in the middle-half. Class 1 has middle-half scores closer together.
D	Class 1 has a better set of scores. Three fourths (3/4) of the scores in Class 1 are above the median for Class 2.

Average Test Scores – Example 2

Sam received the following scores on his math tests:
78, 92, 85, and 97

What score must Sam get on his next math test to have an overall average of 90 ?

Let x = Score on next text = What we're looking for

Average = $\dfrac{Sum}{n}$

$90 = \dfrac{78 + 92 + 85 + 97 + x}{5}$

$90 = \dfrac{351 + x}{5}$

$450 = 351 + x$

$99 = x$

Sam must get a score of 99 on his next math text to have an overall average of 90 (for 5 math tests).

Probability

\multicolumn{2}{c}{**Terms Used When Calculating Probability**}
Permutations $_nP_k$	The number of ways you can pick k items from a group of n items. Here, order matters. Be **P**articular.
Combinations $_nC_k = \binom{n}{k}$	The number of ways you can pick k items from a group of n items. Here, order does NOT matters. It's like grabbing a **C**lump.
Sample Space	The set of all possible outcomes.
Events	The outcome you're looking for.
Probability	The likelihood that the event you're looking for will happen. It is a number between 0 and 1. A probability of 1 means the event is certain to happen. In general, $P = \dfrac{Good\ outcomes}{Possible\ outcomes}$
	For example: The probability of getting a "Heads" when flipping a coin is: $P(H) = \dfrac{1}{2}$

Permutations and Combinations	
Notes	$5! = 5 \cdot 4 \cdot 3 \cdot 2 \cdot 1 = 120$ $0! = 1$ By definition

Permutations $_nP_k$	The number of ways you can pick k items from a group of n items. Here, order matters. Be **P**articular.
	$$_nP_k = \frac{n!}{(n-k)!}$$
	The number of permutations of n objects is $n!$ $_nP_n = n!$

Combinations $_nC_k = \binom{n}{k}$	The number of ways you can pick k items from a group of n items. Here, order does NOT matters. It's like grabbing a **C**lump.
	$$_nC_k = \frac{n!}{k!\,(n-k)!}$$
	The number of combinations of n objects is 1. $_nC_n = 1$

Permutations – Special Case	
Permutations (Special Case) Number of permutations when picking n objects form n items, where: $n = n_1 + n_2 + n_3 + \cdots$	$P = \dfrac{n!}{n_1!\, n_2!\, n_3! \cdots}$

Example	
Find the number of ways the letters in "HUBBUB" can be arranged.	
n	$n = 6$ Total letters $= 6$
n_1	$n_1 = 1$ One H
n_2	$n_2 = 2$ Two U's
n_3	$n_3 = 3$ Three B's
Answer	$P = \dfrac{n!}{n_1!\, n_2!\, n_3!}$ $P = \dfrac{6!}{1!\, 2!\, 3!} = \dfrac{(720)}{(1)(2)(6)} = 60$

Mutually Exclusive and Independent Events	
Consider two sets: $A = \{1, 2, 3, 4\}$ $B = \{3, 4, 5, 6\}$	
Venn Diagram	S A (1 2 (3 4) 5 6) B
Union	$A \cup B = \{1, 2, 3, 4, 5, 6\}$
Intersection	$A \cap B = \{3, 4\}$

Probability of 2 Events, A & B
$P(A \cup B) = P(A) + P(B) - P(A \cap B)$
$P(A \cap B) = P(A) \cdot P(B)$
$\bar{A}$ = Not A = Compliment of A $P(\bar{A}) = 1 - P(A)$
$\bar{B}$ = Not B = Compliment of B $P(\bar{B}) = 1 - P(B)$

	Combinations & Permutations -- Tips
	Combinations and Permutations can be calculated on Desmos (online or app) and on a TI-84 Graphing Calculator as shown below.

Desmos	nCr(5,2) = 10 nPr(5,2) = 20
TI-84 Graphing Calculator	[math], PROB, Select $_nC_r$ or $_nP_r$ 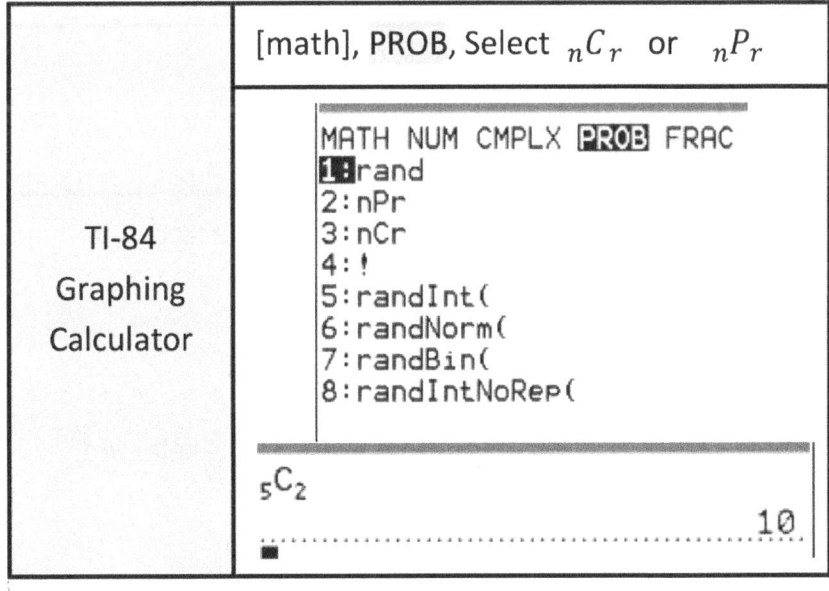

Mutually Exclusive and Independent Events – Ex. 1

Based on past performances during the basketball season, the chances of the top three players (**Allan, Ben, Chad**) making the all-star team are as follows:

$$P(A) = \frac{3}{4} \qquad P(B) = \frac{2}{3} \qquad P(C) = \frac{1}{2}$$

Find the probability of the following events:

A. Allan and Chad are selected for the all-star team, but Ben is not.
B. At least one of the three is selected.

A.	$P(\bar{B}) = 1 - P(B) = 1 - \frac{2}{3} = \frac{1}{3}$ $P = P(A) \cdot P(C) \cdot P(\bar{B})$ $P = \left(\frac{3}{4}\right) \cdot \left(\frac{1}{2}\right) \cdot \left(\frac{1}{3}\right) = \frac{3}{24} = \frac{1}{8}$
B.	$P(Z) = $ Probability none are selected. $P(Z) = P(\bar{A}) \cdot P(\bar{B}) \cdot P(\bar{C})$ $P(Z) = \left(\frac{1}{4}\right) \cdot \left(\frac{1}{3}\right) \cdot \left(\frac{1}{2}\right) = \frac{1}{24}$ $P(\bar{Z}) = $ Probability at least one is selected. $P(\bar{Z}) = 1 - P(Z) = \frac{23}{24}$

Probability – Ex. 2

A box contains four 40W and six 60W light bulbs. If bulbs are selected one by one in random order, What is the probability that at least 2 bulbs must be selected to get one that is 60W?

Given:
$n = 10$

$P(40W) = \dfrac{4}{10} = \dfrac{2}{5}$

$P(60W) = \dfrac{6}{10} = \dfrac{3}{5}$

Think about it: At least 2 selected means you didn't get a 60-W on the first try. So, find the probability of NOT getting a 60W on the first try.

$P(at\ least\ 2) = P(\ not\ getting\ 60W\ on\ first\ try)$

$P(at\ least\ 2) = 1 - P(getting\ 60W\ on\ first\ try)$

$P(at\ least\ 2) = 1 - \dfrac{6}{10} = \dfrac{4}{10} = .40$

Probability – Ex. 3a

On his way to work, a motorist passes through two intersections with traffic lights. He has collected the following information:

$P(A) = .04$	Probability of stopping at 1st traffic signal.
$P(B) = .05$	Probability of stopping at 2nd traffic signal.
$P(A \cup B) = .06$	Probability of stopping at one of the two signals. A or B

What is the probability that he will stop at:
   A. At both signals?
   B. At the first signal but not the second one?
   C. At exactly one signal?

Part A: Probability of stopping at both signals.

$P(A \cap B) = P(A) + P(B) - P(A \cup B)$

$P(A \cap B) = .04 + .05 - .06 = .03$ Answer

Note: $P(A \cup B) = P(A) + P(B) - P(A \cap B)$

Probability – Ex. 3b

On his way to work, a motorist passes through two intersections with traffic lights. He has collected the following information:

$P(A) = .04$	Probability of stopping at 1st traffic signal.
$P(B) = .05$	Probability of stopping at 2nd traffic signal.
$P(A \cup B) = .06$	Probability of stopping at one of the two signals. A or B

What is the probability that he will stop at:
   A. At both signals?
   B. At the first signal but not the second one?
   C. At exactly one signal?

Part B: Probability of stopping at 1st but not 2nd signal.

$= P(stopping\ at\ A) - P(stopping\ at\ A\ and\ B)$
$= P(A) - P(A \cap B)$
$= .04 - .03 = .01$ Answer

Note: $P(A \cap B) = .03$ Found previously.

Probability – Ex. 3c

On his way to work, a motorist passes through two intersections with traffic lights. He has collected the following information:

$P(A) = .04$	Probability of stopping at 1st traffic signal.
$P(B) = .05$	Probability of stopping at 2nd traffic signal.
$P(A \cup B) = .06$	Probability of stopping at one of the two signals. A or B

What is the probability that he will stop at:
A. At both signals?
B. At the first signal but not the second one?
C. At exactly one signal?

Part C: Probability of stopping at exactly one signal.

$P(\text{Just One}) = P(A \cup B) - P(A \cap B)$

$P(\text{Just One}) = .06 - .03 = .03$ Answer

Note: $P(A \cap B) = .03$ Found previously.

Probability – Ex. 3d

On his way to work, a motorist passes through two intersections with traffic lights. He has collected the following information:

$P(A) = .04$	Probability of stopping at 1st traffic signal.
$P(B) = .05$	Probability of stopping at 2nd traffic signal.
$P(A \cup B) = .06$	Probability of stopping at one of the two signals. A or B
$P(A \cap B) = .03$	Probability of stopping at both.

VENN Diagram (Extra)

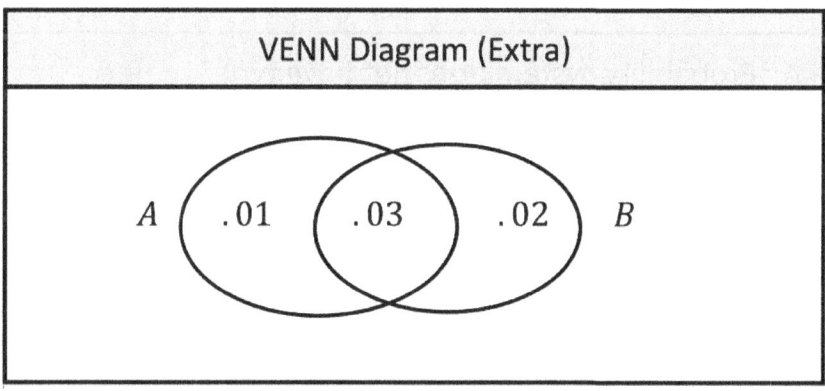

Probability – Ex. 4a

Suppose a system can have three different defects. Let A_i ($i = 1,2,3$) denote the system has defect i.

$P(A_1) = .12$	$P(A_1 \cup A_2) = .13$
$P(A_2) = .07$	$P(A_1 \cup A_3) = .14$
$P(A_3) = .05$	$P(A_2 \cup A_3) = .10$
	$P(A_1 \cap A_2 \cap A_3) = .01$

What is the probability that the system:
A. Does NOT have a type 1 defect?
B. Has both type 1 and type 2 defects?
C. Has both type 1 and type 2 but NOT type 3 defects?
D. Has at most two of these defects?

A: Probability system does not have type 1 defect.

$P(\bar{A}) = P(not\ A_1)$

$P(\bar{A}) = 1 - P(A_1)$

$P(\bar{A}) = 1 - .12 = .87$ Answer

Probability – Ex. 4b

Suppose a system can have three different defects. Let A_i ($i = 1,2,3$) denote the system has defect i.

$P(A_1) = .12$	$P(A_1 \cup A_2) = .13$
$P(A_2) = .07$	$P(A_1 \cup A_3) = .14$
$P(A_3) = .05$	$P(A_2 \cup A_3) = .10$
	$P(A_1 \cap A_2 \cap A_3) = .01$

What is the probability that the system:
A. Does NOT have a type 1 defect?
B. Has both type 1 and type 2 defects?
C. Has both type 1 and type 2 but NOT type 3 defects?
D. Has at most two of these defects?

B: Probability system has type 1 and type 2 defects?

$P(A_1 \cap A_2) = P(A_1 \text{ and } A_2)$

$P(A_1 \cap A_2) = P(A_1) + P(A_2) - P(A_1 \cup A_2)$

$P(A_1 \cap A_2) = .12 + .07 - .13 = .06$ Answer

Probability – Ex. 4c

Suppose a system can have three different defects. Let A_i ($i = 1,2,3$) denote the system has defect i.

$P(A_1) = .12$	$P(A_1 \cup A_2) = .13$
$P(A_2) = .07$	$P(A_1 \cup A_3) = .14$
$P(A_3) = .05$	$P(A_2 \cup A_3) = .10$
$P(A_1 \cap A_2) = .06$	$P(A_1 \cap A_2 \cap A_3) = .01$

What is the probability that the system:
A. Does NOT have a type 1 defect?
B. Has both type 1 and type 2 defects?
C. Has both type 1 and type 2 but NOT type 3 defects?
D. Has at most two of these defects?

C: Probability system has type 1 and type 2 defects but NOT type 3 defects?

$P(A_1 \cap A_2 \cap \overline{A_3}) = P(A_1 \cap A_2) - P(A_1 \cap A_2 \cap A_3)$

$P(A_1 \cap A_2 \cap \overline{A_3}) = .06 - .01 = .05$ Answer

Venn Diagram	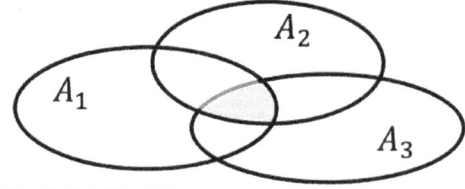

Probability – Ex. 4d

Suppose a system can have three different defects. Let A_i ($i = 1,2,3$) denote the system has defect i.

$P(A_1) = .12$	$P(A_1 \cup A_2) = .13$
$P(A_2) = .07$	$P(A_1 \cup A_3) = .14$
$P(A_3) = .05$	$P(A_2 \cup A_3) = .10$
$P(A_1 \cap A_2) = .06$	$P(A_1 \cap A_2 \cap A_3) = .01$

What is the probability that the system:
A. Does NOT have a type 1 defect?
B. Has both type 1 and type 2 defects?
C. Has both type 1 and type 2 but NOT type 3 defects?
D. Has at most two of these defects?

D: Probability system has at most two defects?

$P(at\ most\ 2) = 1 - P(all\ 3\ defects)$
$P(at\ most\ 2) = 1 - P(A_1 \cap A_2 \cap A_3)$
$P(at\ most\ 2) = 1 - .01 = .09$ Answer

Venn Diagram	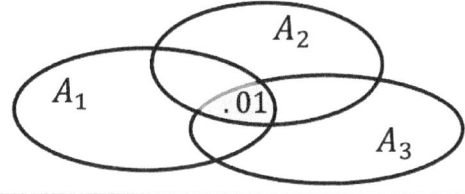

Probability – Ex. 4e.

Suppose a system can have three different defects. Let A_i ($i = 1,2,3$) denote the system has defect i.

$P(A_1) = .12$	$P(A_1 \cup A_2) = .13$
$P(A_2) = .07$	$P(A_1 \cup A_3) = .14$
$P(A_3) = .05$	$P(A_2 \cup A_3) = .10$
$P(A_1 \cap A_2) = .06$	$P(A_1 \cap A_2 \cap A_3) = .01$

Venn Diagram (Extra)

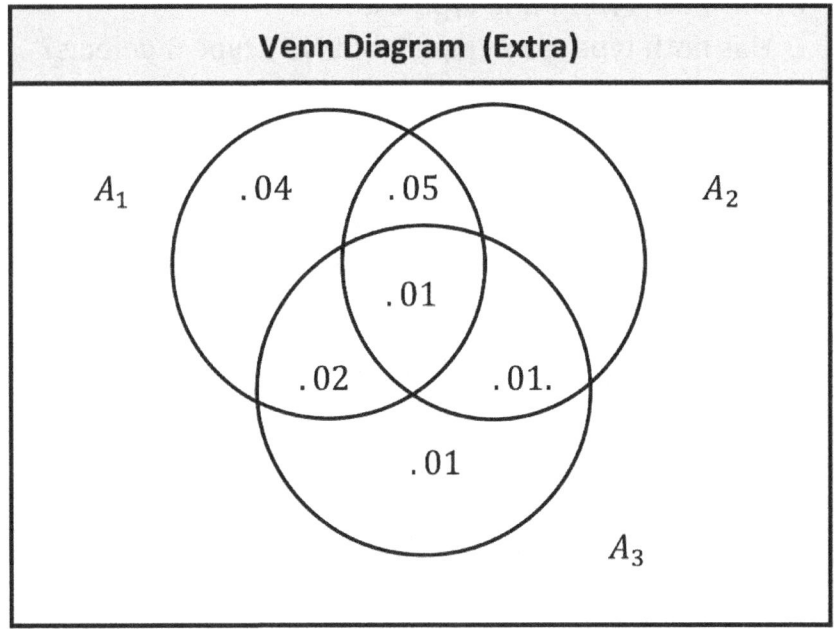

Limits

Definition of a Limit

$$\lim_{x \to a} f(x) = L$$

The limit of $f(x)$, as x approaches a, equals L

As x approaches a, <u>from either the left or right</u>, The function approaches a value of L.

In other words ...
As x gets close to a from either side, the function gets close to L. The function may or may not be defined at $f(a)$. That's why we just get close to it.

Left-Hand Limit	**Right-Hand Limit**
$\lim_{x \to a^-} f(x) = L$	$\lim_{x \to a^+} f(x) = L$

IFF $\lim_{x \to a^+} f(x) = L$ <u>and</u> $\lim_{x \to a^-} f(x) = L$

THEN $\lim_{x \to a} f(x) = L$

If both the left and right limits are the same we can simply say ...

When x approaches a
the function approaches L.

Limit Laws
$\lim_{x \to a} f(x) = f(a)$ direct substitution
$\lim_{x \to a} [f(x) \pm g(x)] = \lim_{x \to a} f(x) \pm \lim_{x \to a} g(x)$
$\lim_{x \to a} [f(x) \cdot g(x)] = \lim_{x \to a} f(x) \cdot \lim_{x \to a} g(x)$
$\lim_{x \to a} [c \cdot f(x)] = c \cdot \lim_{x \to a} f(x)$
$\lim_{x \to a} \left[\dfrac{f(x)}{g(x)} \right] = \dfrac{\lim_{x \to a} f(x)}{\lim_{x \to a} g(x)}$
$\lim_{x \to a} [f(x)]^n = \left[\lim_{x \to a} f(x) \right]^n$

---- Limits at Horizontal Asymptotes (HA) ----

The line $y = L$ is a horizontal asymptote of $f(x)$ if either: $\lim\limits_{x \to \infty} f(x) = L$ or $\lim\limits_{x \to -\infty} f(x) = L$

Limits at infinity are horizontal asymptotes.
They represent the "far end" behavior of the function.

Note: Some functions may cross their horizontal asymptote but eventually approach the HA and settle down, when x is very large.

Limits at Vertical Asymptotes (VA)

$x = a$ is a Vertical Asymptote (VA) of a function:
$$y = f(x)$$
If at least one of the following is true:

$\lim\limits_{x \to a} f(x) = \infty$	$\lim\limits_{x \to a} f(x) = -\infty$
$\lim\limits_{x \to a^+} f(x) = \infty$	$\lim\limits_{x \to a^+} f(x) = -\infty$
$\lim\limits_{x \to a^-} f(x) = \infty$	$\lim\limits_{x \to a^-} f(x) = -\infty$

In other words ...
    Infinite limits occur at vertical asymptotes.

Infinite Limit
$$\lim_{x \to a} f(x) = \infty$$ The limit of $f(x)$, as x approaches a, from either side, becomes extremely large.
In other words ... As x gets close to the value of a, from either side, the function becomes very, very large. Note: ∞ is not a number. Since ∞ is not a number, there is actually no limit. The limit does NOT exist! The function is not limited to a particular number. It's just large.

--- Limits at Horizontal Asymptotes (HA) ---
Some Examples

$f(x) = \dfrac{(x-1)}{(x^2+1)}$ $\displaystyle\lim_{x \to \pm\infty} f(x) = 0$	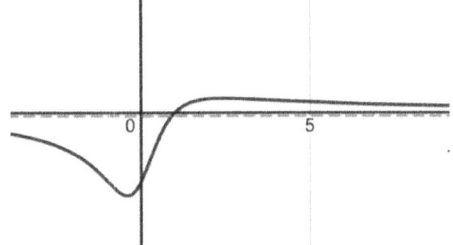
$f(x) = \dfrac{1}{x} + 2$ $\displaystyle\lim_{x \to \pm\infty} f(x) = 2$	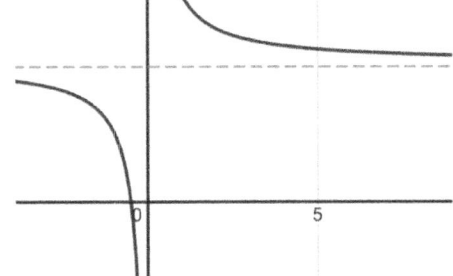
$f(x) = \sin\left(\dfrac{1}{x}\right)$ $\displaystyle\lim_{x \to \pm\infty} f(x) = 0$	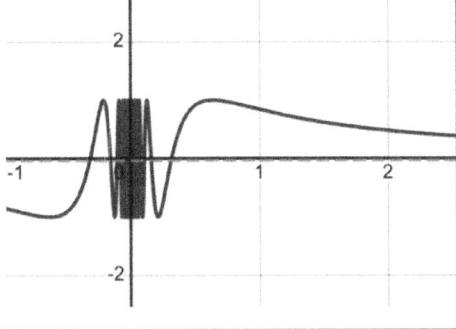

Limits at Horizontal Asymptotes (HA)
ArcTan Example: $y = Tan^{-1} x$

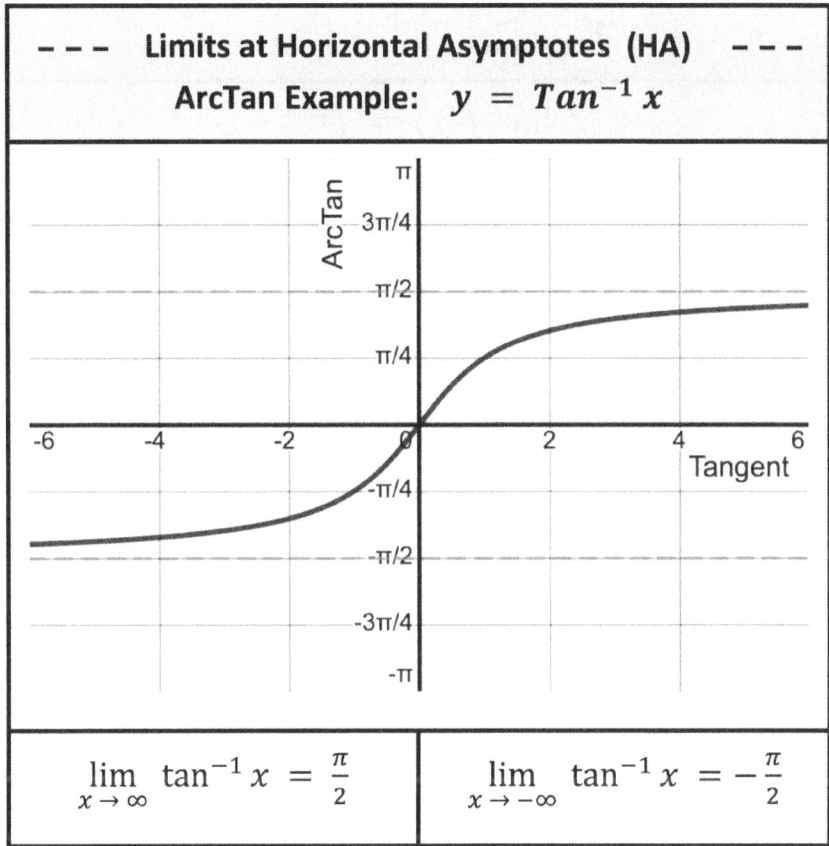

$$\lim_{x \to \infty} \tan^{-1} x = \frac{\pi}{2}$$

$$\lim_{x \to -\infty} \tan^{-1} x = -\frac{\pi}{2}$$

Continuous Function at a	
$\lim_{x \to a} f(x) = f(a)$	
$\lim_{x \to a^+} f(x) = f(a)$	Continuous from the right
$\lim_{x \to a^-} f(x) = f(a)$	Continuous from the left

In other words …
- As x approaches a, from either side, the function approaches $f(a)$.
- There are NO gaps or jumps.

Notes
- Polynomial functions are continuous everywhere.
- Rational functions are continuous where defined.

Intermediate Value Theorem (IVT)

If $f(x)$ is continuous on $[a, b]$

Where $f(a) < N < f(b)$

Then, there is a number c in (a, b)

Such that $f(c) = N$.

Between two known values of a continuous function, the function will have another value equal to or between those two values.

IVT Example

Given: $f(x) = x^3 - 6x + 2$
Show there is a root between 2 and 3.

Continuous	All polynomials are continuous.
Check Boundaries	$f(2) = (2)^3 - 6(2) + 2 = -2$ $f(3) = (3)^3 - 6(3) + 2 = 11$
$-2 < 0 < 11$	A root is in interval $(2, 3)$ Because there is a sign change.

Limits – Examples	
Limit	**Answer**
$\lim\limits_{x \to 5} 3x^2$	$= 3(5)^2 = 75$
$\lim\limits_{x \to 3} \dfrac{x-3}{x^2-9}$	$= \lim\limits_{x \to 3} \dfrac{(x-3)}{(x-3)(x+3)}$ $= \lim\limits_{x \to 3} \dfrac{1}{(x+3)} = \dfrac{1}{6}$
$\lim\limits_{x \to 0} \dfrac{5x^3 + 4x^2}{x^2}$	$= \lim\limits_{x \to 0} \dfrac{5x + 4}{1}$ $= \lim\limits_{x \to 0} 5x + 4 = 4$
$\lim\limits_{x \to 0} \left(x^2 + \dfrac{\cos 3x}{100}\right)$	$= 0 + \dfrac{1}{100} = 0.01$
$\lim\limits_{x \to 5} 12$	$= 12$ Note: Here, $f(x) = 12$ So, $f(5) = 12$ The function is always 12.
$\lim\limits_{x \to 3} \dfrac{x+3}{x^2-9}$	$= \lim\limits_{x \to 3} \dfrac{(x+3)}{(x-3)(x+3)}$ $= \lim\limits_{x \to 3} \dfrac{1}{(x-3)} = \dfrac{1}{0} = \infty$ DNE

Limits – More Examples	
Limit	Answer
$\lim\limits_{x \to 1} \dfrac{1 - \frac{1}{x}}{x - 1}$	$= \lim\limits_{x \to 1} \dfrac{\frac{1}{x}(x - 1)}{(x - 1)}$ $= \lim\limits_{x \to 1} \dfrac{1}{x} = \dfrac{1}{1} = 1$
$\lim\limits_{x \to -1} \dfrac{1 - \frac{1}{x}}{x - 1}$	$= \dfrac{1 + 1}{-1 - 1}$ $= \dfrac{2}{-2} = -1$
$\lim\limits_{x \to 3} \dfrac{x^2 + 2x - 15}{x^2 - 9}$	$= \lim\limits_{x \to 3} \dfrac{(x - 3)(x + 5)}{(x - 3)(x + 3)}$ $= \lim\limits_{x \to 3} \dfrac{(x + 5)}{(x + 3)}$ $= \dfrac{8}{6} = \dfrac{4}{3}$

One Sided Limits – Examples

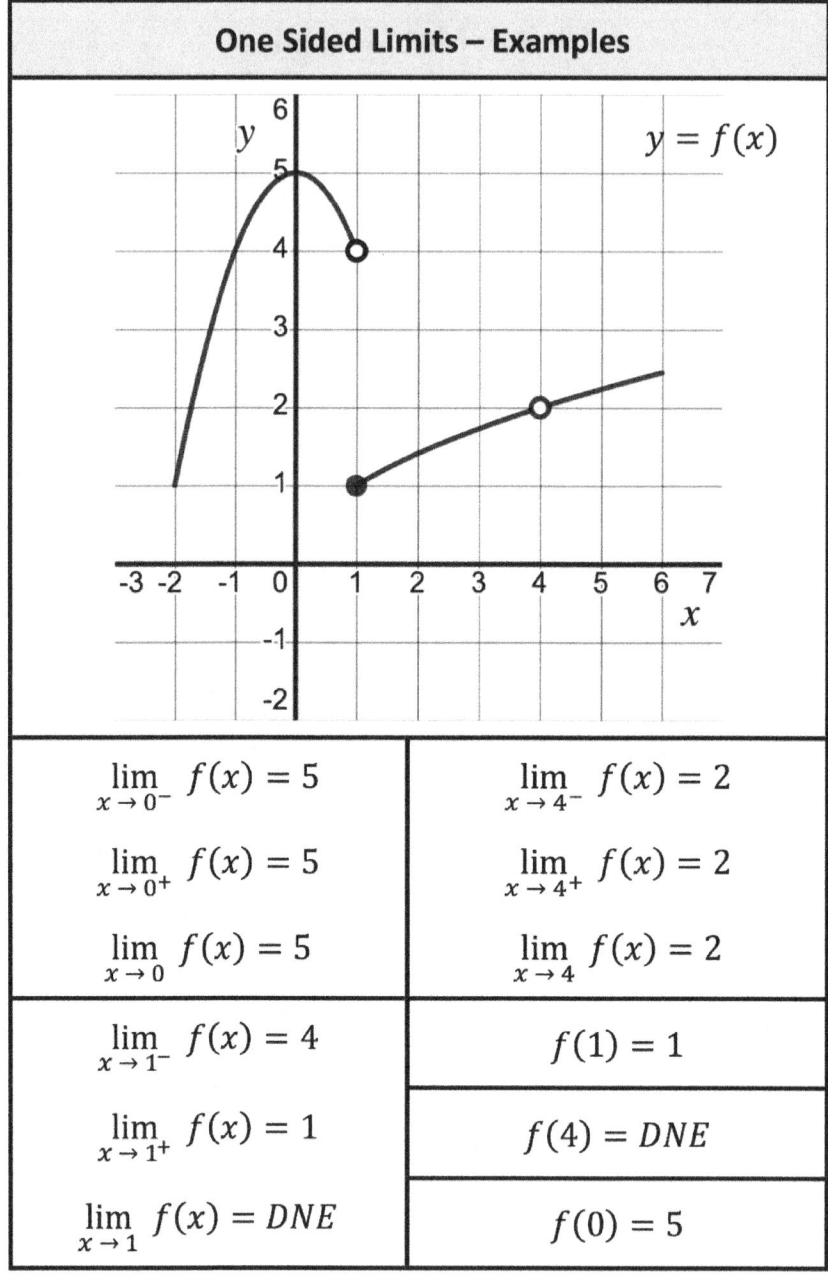

$\lim_{x \to 0^-} f(x) = 5$	$\lim_{x \to 4^-} f(x) = 2$
$\lim_{x \to 0^+} f(x) = 5$	$\lim_{x \to 4^+} f(x) = 2$
$\lim_{x \to 0} f(x) = 5$	$\lim_{x \to 4} f(x) = 2$
$\lim_{x \to 1^-} f(x) = 4$	$f(1) = 1$
$\lim_{x \to 1^+} f(x) = 1$	$f(4) = DNE$
$\lim_{x \to 1} f(x) = DNE$	$f(0) = 5$

Useful Equations

When evaluating limits, there are some simple limits that are useful to know. Later, you will learn techniques (e.g. L'Hospital's Rule) that will help with evaluating limits.

For now, just memorize these limits to help with evaluating a variety of limits.

Memorize These Special Limits

$$\lim_{x \to 0} \frac{\sin x}{x} = 1$$

$$\lim_{x \to 0} \frac{1 - \cos x}{x} = 0$$

$$\lim_{x \to \infty} \frac{c}{x} = 0$$

Limits – Ex. 1	
Evaluate:	$\lim\limits_{x \to \infty} \dfrac{x^2 + 5}{7 - 3x + 2x^3}$

Divide numerator and denominator by x^3	$\lim\limits_{x \to \infty} \dfrac{x^2 + 5}{7 - 3x + 2x^3}$ $\lim\limits_{x \to \infty} \dfrac{\dfrac{1}{x} + \dfrac{5}{x^3}}{\dfrac{7}{x^3} - \dfrac{3}{x^2} + 2}$
Substitute $x = \infty$ Then evaluate	$\dfrac{0 + 0}{0 - 0 + 2} = \dfrac{0}{2} = 0$

Limits – Ex. 2
Evaluate: $\lim\limits_{x \to 5} f(x)$
Given: $f(x) = \begin{cases} 3x+2 & , x > 5 \\ 1 & , x = 5 \\ x^2 + 8 & , x < 5 \end{cases}$

Evaluate limit From the left.	$\lim\limits_{x \to 5^-} f(x) = 5^2 + 8$ $\lim\limits_{x \to 5^-} f(x) = 33$
Evaluate limit from the right.	$\lim\limits_{x \to 5^+} f(x) = 3(5) + 2$ $\lim\limits_{x \to 5^+} f(x) = 17$
Limit exists only if limits from left and right are equal.	$\lim\limits_{x \to 5} f(x) = DNE$ $DNE =$ Does Not Exist

	Limits – Ex. 3	
Evaluate:	$\lim\limits_{x \to 0} \dfrac{\tan(2x)}{x}$	
Recall:	$\lim\limits_{x \to 0} \dfrac{\sin x}{x} = 1$	

Also, recall Trig. Double angles	$\sin(2u) = 2 \sin u \cdot \cos u$ $\cos(2u) = \cos^2 u - \sin^2 u$
Simplify. Use Trig. Substitutions	$\lim\limits_{x \to 0} \dfrac{1}{x} \left[\dfrac{\sin(2x)}{\cos(2x)} \right]$ $\lim\limits_{x \to 0} \dfrac{1}{x} \left[\dfrac{2 \sin x \cdot \cos x}{\cos^2 x - \sin^2 x} \right]$
Isolate the known limit.	$\lim\limits_{x \to 0} \left(\dfrac{\sin x}{x} \right) \cdot \left[\dfrac{2 \cdot \cos x}{\cos^2 x - \sin^2 x} \right]$
Substitute $x = 0$ Then evaluate	$(1) \left[\dfrac{2 \cdot (1)}{1 - 0} \right] = \dfrac{2}{1} = 2$

Limits – Ex. 4
Evaluate: $\displaystyle\lim_{x \to 0} \frac{1 - \cos x + 8x}{x}$
Recall: $\displaystyle\lim_{x \to 0} \frac{1 - \cos x}{x} = 0$

Rearrange	$\displaystyle\lim_{x \to 0} \left[\frac{1 - \cos x}{x} + \frac{8x}{x} \right]$
Simplify	$\displaystyle\lim_{x \to 0} \left[\frac{1 - \cos x}{x} + 8 \right]$
Evaluate	$\displaystyle\lim_{x \to 0} \left(\frac{1 - \cos x}{x} \right) + \lim_{x \to 0} (8)$ $0 + 8 = 8$

Limits – Ex. 5
Evaluate: $\lim\limits_{x \to 0} \dfrac{|x|}{x}$

Evaluate from the left. Note: x is negative	$\lim\limits_{x \to 0} \dfrac{|x|}{x}$ $\dfrac{positive}{negative} = -1$
Evaluate from the right Note: x is positive	$\lim\limits_{x \to 0} \dfrac{|x|}{x} = \dfrac{positive}{positive} = 1$
A limit exists only if the limit from the left & right are equal.	$\lim\limits_{x \to 0} \dfrac{|x|}{x} = DNE$ $DNE =$ Does Not Exist

Limits – Ex. 6
Evaluate: $\displaystyle\lim_{x \to 3^-} \frac{x+1}{x-3}$

Evaluate this limit from the left for $x \approx 2.9$ 2.9 is a little less than 3.	$\dfrac{3+1}{2.9-3}$ $\dfrac{4}{negative\ zero} = -\infty$

Note: When looking at the original limit, it is easy to see it is in the form: $\dfrac{c}{0} = \pm\infty$.

For this one-sided limit, denominator is negative as it approaches zero.

Limits – Ex. 7

Evaluate: $\lim_{x \to 3} f(g(x))$

Given: $f(x)$ and $g(x)$ are both continuous functions with the following table of values.

x	1	2	3	4
$f(x)$	3	6	9	12
$g(x)$	0	1	2	3

Work from the inside to the outside.	$\lim_{x \to 3} f(g(x))$ $f\left(\lim_{x \to 3} g(x)\right)$ $f(g(3))$ $f(2) = 6$

Limits – Ex. 8

Given the graph of $f(x)$, evaluate the limits.

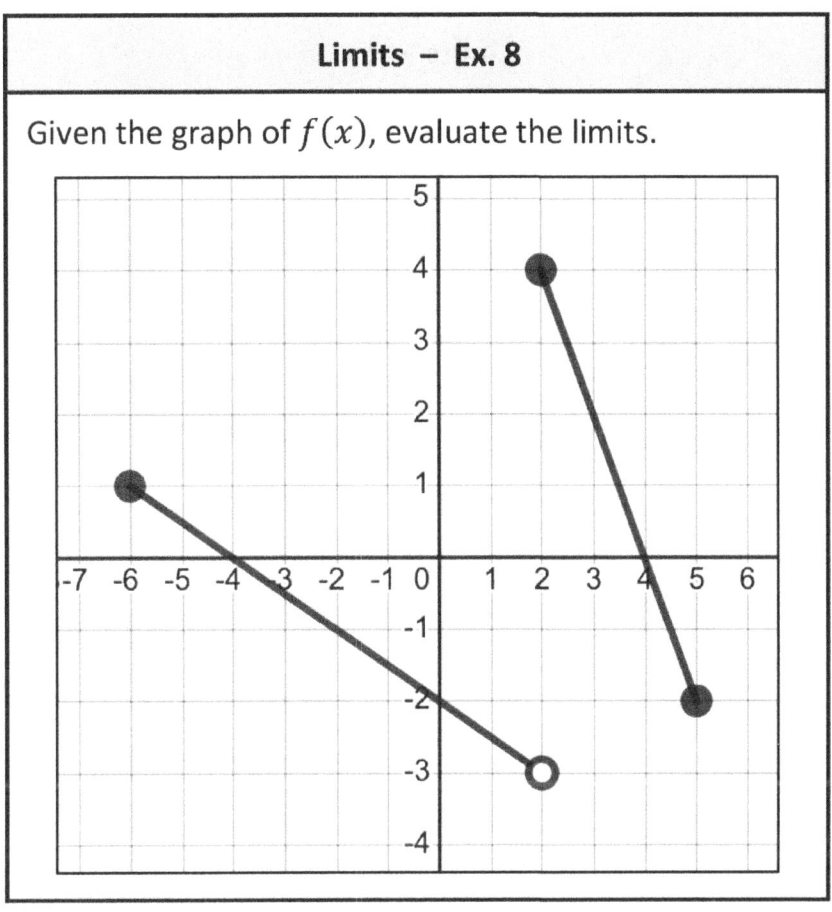

$\lim\limits_{x \to 2^-} f(x)$	$\lim\limits_{x \to 2^-} f(x) = -3$
$\lim\limits_{x \to 2^+} f(x)$	$\lim\limits_{x \to 2^+} f(x) = 4$
$\lim\limits_{x \to 2} f(x)$	$\lim\limits_{x \to 2} f(x) = DNE$

Limits – Ex. 9

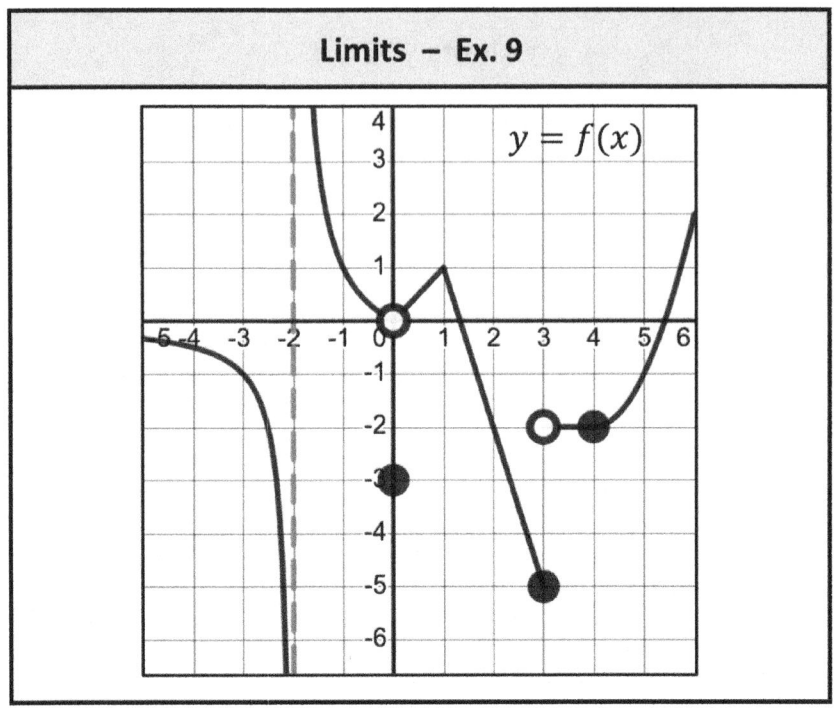

$\lim\limits_{x \to 2^+} f(x)$	$\lim\limits_{x \to 2^+} f(x) = \infty$
$f(0)$	$f(0) = -3$
$\lim\limits_{x \to 0^-} f(x)$	$\lim\limits_{x \to 0^-} f(x) = 0$
$\lim\limits_{x \to 3^-} f(x)$	$\lim\limits_{x \to 3^-} f(x) = -5$
$\lim\limits_{x \to 3} f(x)$	$\lim\limits_{x \to 3} f(x) = DNE$
$\lim\limits_{x \to -\infty} f(x)$	$\lim\limits_{x \to -\infty} f(x) = 0$

Limits – Ex. 10

Given: $\lim\limits_{x \to a} f(x) = 20$ and $\lim\limits_{x \to a} g(x) = -5$

Evaluate: The following limits.

$\lim\limits_{x \to a} [g(x) - f(x)]$	$\lim\limits_{x \to a} g(x) - \lim\limits_{x \to a} f(x)$ $-5 - 20 = -25$
$\lim\limits_{x \to a} [2f(x) + g(x)]$	$2 \lim\limits_{x \to a} f(x) + \lim\limits_{x \to a} g(x)$ $2(20) + (-5)$ $40 - 5 = 35$
$\lim\limits_{x \to a} [g(x) \cdot f(x)]$	$\left[\lim\limits_{x \to a} g(x) \right] \cdot \left[\lim\limits_{x \to a} f(x) \right]$ $(-5) \cdot (20) = -100$

Limits – Ex. 11
Evaluate: $\lim\limits_{x \to -2} \dfrac{x^2 + 5x + 6}{x^2 - 5x - 10}$

Factor the numerator and denominator.	$\lim\limits_{x \to -2} \dfrac{(x+2)(x+3)}{(x+2)(x-5)}$
Cancel terms	$\lim\limits_{x \to -2} \dfrac{(x+3)}{(x-5)}$
Use substitution to evaluate.	$\dfrac{(-2+3)}{(-2-5)} = -\dfrac{1}{7}$

Limits – Ex. 12
Evaluate: $\lim\limits_{x \to 1} \dfrac{\sqrt{3x+2} - \sqrt{5}}{x-1}$

Eliminate radicals in the numerator.	$\lim\limits_{x \to 1} \dfrac{\sqrt{3x+2} - \sqrt{5}}{x-1} \cdot \left[\dfrac{\sqrt{3x+2} + \sqrt{5}}{\sqrt{3x+2} + \sqrt{5}} \right]$ $\lim\limits_{x \to 1} \dfrac{(3x+2) - (5)}{(x-1) \cdot (\sqrt{3x+2} + \sqrt{5})}$ $\lim\limits_{x \to 1} \dfrac{3x - 3}{(x-1) \cdot (\sqrt{3x+2} + \sqrt{5})}$
Simplify	$\lim\limits_{x \to 1} \dfrac{3(x-1)}{(x-1) \cdot (\sqrt{3x+2} + \sqrt{5})}$ $\lim\limits_{x \to 1} \dfrac{3}{(\sqrt{3x+2} + \sqrt{5})}$
Use substitution to evaluate.	$\dfrac{3}{\left(\sqrt{3(1)+2} + \sqrt{5}\right)}$ $\dfrac{3}{(\sqrt{5} + \sqrt{5})} = \dfrac{3}{2\sqrt{5}}$

Limits – Ex. 13

Evaluate: $\displaystyle\lim_{x \to 0} \frac{\frac{1}{x+5} - \frac{1}{5}}{x}$

Write numerator as a single fraction.	$\displaystyle\lim_{x \to 0} \frac{\frac{1}{x+5}\left(\frac{5}{5}\right) - \frac{1}{5}\left(\frac{x+5}{x+5}\right)}{x}$ $\displaystyle\lim_{x \to 0} \frac{\left(\frac{5 - (x+5)}{5(x+5)}\right)}{x}$	
Then simplify	$\displaystyle\lim_{x \to 0} \frac{\left(\frac{-x}{5(x+5)}\right)}{x}$ $\displaystyle\lim_{x \to 0} \frac{-x}{5x(x+5)}$ $\displaystyle\lim_{x \to 0} \frac{-1}{5(x+5)}$	
Evaluate	$\dfrac{-1}{5(0+5)} = \dfrac{-1}{5(5)} = -\dfrac{1}{25}$	

Limits – Ex. 14

Given: $5x - 4 \leq f(x) \leq x^2 + 2$

Find: $\lim_{x \to 2} f(x)$ and Justify your answer.

Use the Squeeze Theorem:
If $a \leq b \leq c$ and $a = c$ Then: $a = b = c$

$5x - 4 \leq f(x) \leq x^2 + 2$

$\lim_{x \to 2} (5x - 4) \leq \lim_{x \to 2} f(x) \leq \lim_{x \to 2} (x^2 + 2)$

$(5(2) - 4) \leq \lim_{x \to 2} f(x) \leq (2^2 + 2)$

$(10 - 4) \leq \lim_{x \to 2} f(x) \leq (4 + 2)$

$6 \leq \lim_{x \to 2} f(x) \leq 6$

Conclusion:

$\lim_{x \to 2} f(x) = 6$ By the Squeeze Theorem

Limits – Ex. 15
Given: $f(x) = \begin{cases} x^2 + k &, x \leq 0 \\ 5 - \sin x &, x > 0 \end{cases}$ Find: k that will make $f(x)$ continuous at $x = 0$. Use definition of continuity to justify your answer.

	$x^2 + k \;=\; 5 - \sin x$ $\lim_{x \to 0^-} (x^2 + k) \;=\; \lim_{x \to 0^+} (5 - \sin x)$ $(0 + k) \;=\; (5 - 0)$ $k \;=\; 5$
Recall the 3 conditions of continuity at $x = a$.	$\lim_{x \to a^-} f(x) \text{ exists}$ $\lim_{x \to a^+} f(x) \text{ exists}$ $\lim_{x \to a} f(x) = f(a)$
Three conditions of continuity at $x = a$.	$\lim_{x \to 0^-} f(x) = 0^2 + 5 = 5$ $\lim_{x \to 0^+} f(x) = 5 - \sin 0 = 5$ $f(0) = 0^2 + 5 = 5$ So, $f(x)$ is continuous at $x = 0$

	Limits – Ex. 16
	Given: $f(x) = x^2 - 2x - 1$ on $[-2, 10]$ Does the Intermediate Value Theorem (IVT) Guarantee there is a value $x = c$ such that $f(c) = 23$? Justify your answer.

Recall, The IVT	If $f(x)$ is continuous on $[a, b]$ Where $f(a) < N < f(b)$ Then there is a number, c, in (a, b) where $f(c) = N$
Continuous?	$f(x)$ is a polynomial so it is cont.
Evaluate $f(x)$ at the boundaries.	$f(-2) = (-2)^2 - 2(-2) - 1 = 7$ $f(10) = (10)^2 - 2(10) - 1 = 79$ $7 < N < 79$ True when $N = 23$
Find c where $f(c) = 23$	$c^2 - 2c - 1 \;\;= 23$ $c^2 - 2c - 24 \;= 0$ $(c - 6)(c + 4) = 0$ $c = 6, -4$ -4 not in interval. $c = 6$

Difference Quotient (Derivative)

The Difference Quotient

The **difference quotient** is used to find the slope of a secant line, near the point $(x, f(x))$.

$$\text{slope} = \frac{\text{Change in } y}{\text{Change in } x} = \frac{f(x+h) - f(x)}{(x+h) - x}$$

$$\text{slope} = \frac{f(x+h) - f(x)}{h}$$

If $x = a$

$$\text{slope} = \frac{f(a+h) - f(a)}{h}$$

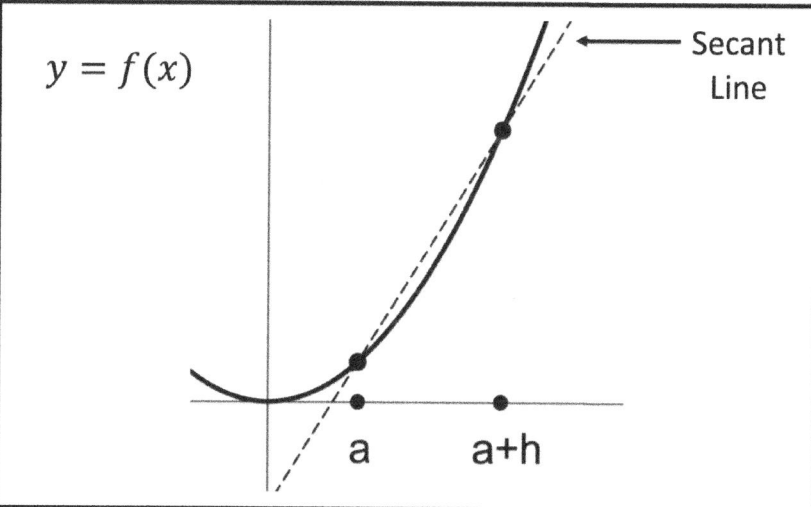

If h is very small, then the secant line is almost a tangent line.

The Difference Quotient
The **difference quotient** is used to find the slope of a secant line, near the point $(x, f(x))$.
$$slope = \frac{f(x+h) - f(x)}{h}$$

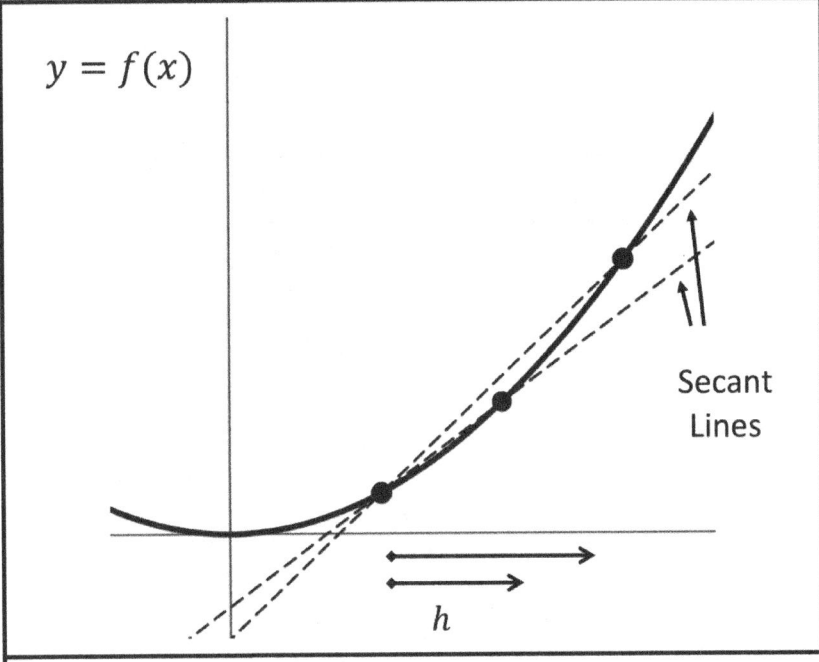

As h gets smaller,
the secant line is more like a tangent line.

The Difference Quotient
Find the Slope of a Secant Line -- $h = 1$

Given: $f(x) = x^2 + 5$

Use the difference quotient to find the slope of a secant line near $x = 2$. Use $h = 1$

$$slope = \frac{f(a+h) - f(a)}{h}$$

$$slope = \frac{f(2+1) - f(2)}{1} = \frac{f(3) - f(2)}{1}$$

$$slope = \frac{[(3)^2 + 5] - [(2)^2 + 5]}{1}$$

$$slope = \frac{(3)^2 + 5 - (2)^2 - 5}{1}$$

$$slope = \frac{(3)^2 - (2)^2}{1} = 9 - 4 = 5$$

The Difference Quotient
Find the Slope of a Secant Line -- $h = h$

Given: $f(x) = x^2 + 5$
Use the difference quotient to find the slope of a secant line near $x = 2$. Use $h = h$

$$slope = \frac{f(a+h) - f(a)}{h}$$

$$slope = \frac{f(2+h) - f(2)}{h} = \frac{f(2+h) - f(2)}{1}$$

$$slope = \frac{[(2+h)^2 + 5] - [(2)^2 + 5]}{h}$$

$$slope = \frac{[2^2 + 4h + h^2 + 5] - [2^2 + 5]}{h}$$

$$slope = \frac{4h + h^2}{h} = 4 + h$$

| **Derivative** | $=\dfrac{Change\ in\ y}{Change\ in\ x}$ | $=\dfrac{\Delta y}{\Delta x}$ | $=\dfrac{dy}{dx}$ |

The **derivative** is the instantaneous **rate of change** of a function. It is the slope of the tangent line to the graph of a function, at a point $(x, f(x))$.

$$Derivative = \lim_{h \to 0} \frac{f(x+h) - f(x)}{h}$$

If $x = a$

$$Derivative = \lim_{h \to 0} \frac{f(a+h) - f(a)}{h}$$

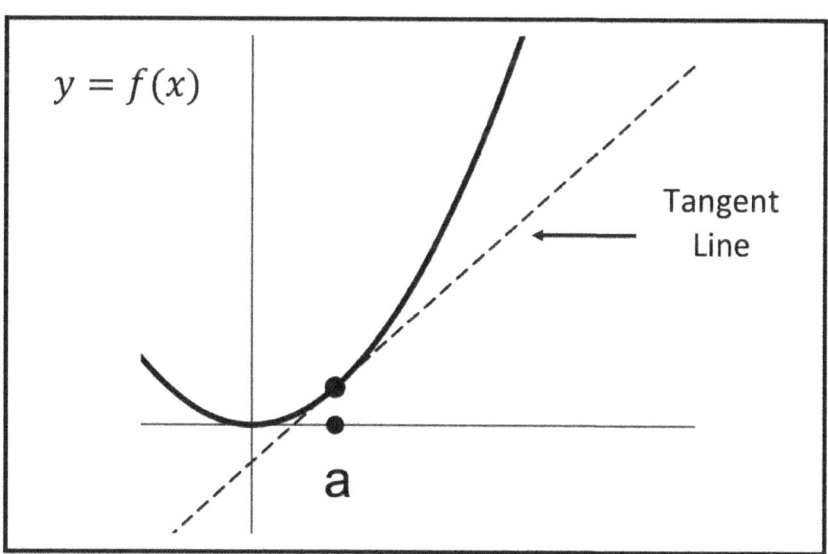

Derivative at $x = a$
Given: $f(x) = x^2 + 5$ Use the limit definition of a derivative to find the slope of the tangent line at $x = 2$.

$Slope = \lim_{h \to 0} \dfrac{f(x+h) - f(x)}{h}$
$Slope = \lim_{h \to 0} \dfrac{f(2+h) - f(2)}{h}$
$Slope = \lim_{h \to 0} \dfrac{[(2+h)^2 + 5] - [(2)^2 + 5]}{h}$
$Slope = \lim_{h \to 0} \dfrac{[4 + 4h + h^2 + 5] - [4 + 5]}{h}$
$Slope = \lim_{h \to 0} \dfrac{4h + h^2}{h} = \lim_{h \to 0} 4 + h = 4$
$Slope = \dfrac{dy}{dx} = 4$, at $x = 2$

Derivative at $x = x$

Given: $f(x) = x^2 + 5$

Use the limit definition of a derivative to find the slope of the tangent line for any x. $\quad x = x$.

$$Slope = \lim_{h \to 0} \frac{f(x+h) - f(x)}{h}$$

$$Slope = \lim_{h \to 0} \frac{[(x+h)^2 + 5] - [(x)^2 + 5]}{h}$$

$$Slope = \lim_{h \to 0} \frac{[x^2 + 2xh + h^2 + 5] - [x^2 + 5]}{h}$$

$$Slope = \lim_{h \to 0} \frac{2xh + h^2}{h} = \lim_{h \to 0} (2x + h)$$

$$Slope = \frac{dy}{dx} = 2x \qquad \text{The slope at any } x$$

When $x = 2$, $\dfrac{dy}{dx} = 2(2) = 4$

This is what was found, previously.

Derivative Notation for $y = f(x)$	
The derivative of a function $f(x)$ has several notations as listed below.	

$f'(x)$	Lagrange's Notation (pronounced "f prime"
f'	
$\dfrac{d}{dx} f(x)$	Leibniz's Notation
$\dfrac{dy}{dx}$	
$\dot{f}$	Newton's Notation
$\dot{y}$	

The triangle (Δ) is a Greek symbol called "delta" and means change.

You may see: $\dfrac{y_2 - y_1}{x_2 - x_1} = \dfrac{\Delta y}{\Delta x} = \dfrac{dy}{dx}$

Equation of Tangent Line – Ex. 1a

Given: $f(x) = x^3$ Use the limit definition of a derivative to find the slope of the tangent line at $x = 2$. Then, find the equation of the tangent line.

$Slope = \lim_{h \to 0} \dfrac{f(x+h) - f(x)}{h}$

$m = \lim_{h \to 0} \dfrac{[(x+h)^3] - [(x)^3]}{h}$

$m = \lim_{h \to 0} \dfrac{[(x+h)(x^2 + 2xh + h^2)] - [x^3]}{h}$

$m = \lim_{h \to 0} \dfrac{[x^3 + 2x^2h + xh^2 + hx^2 + 2xh^2 + h^3] - [x^3]}{h}$

$m = \lim_{h \to 0} \dfrac{3x^2h + 3xh^2 + h^3}{h}$

$m = \lim_{h \to 0} (3x^2 + 3xh + h^2)$

$m = 3x^2$

When $x = 2$,
Slope of tangent line $= m = 3(2)^2 = 12$

	Equation of Tangent Line – Ex. 1b
	Given: $f(x) = x^3$ Use the limit definition of a derivative to find the slope of the tangent line at $x = 2$. Then, find the equation of the tangent line.

Previously Found	$m = 12$
Find point on the curve when $x = 2$	$f(2) = (2)^3 = 8$ Point: $(x, y) = (2, 8)$
Find equation of a straight line through point $(2,8)$ with slope $m = 12$	$y = mx + b$ $y = 12x + b$ Use point $(x, y) = (2, 8)$ $8 = 12(2) + b$ $b = -16$
Tangent line Eqn.	$y = 12x - 16$
Graph (Extra)	

Equation of Tangent Line – Ex. 2

Given: $f(x) = x^3$ Use the limit definition of a derivative to find the slope of the tangent line at x. Then, find two points on $f(x)$, where the slope of the tangent line is 6.

See Previous Example	$m = \lim\limits_{h \to 0} \dfrac{[(x+h)^3] - [(x)^3]}{h}$	
	$m = 3x^2$	
Find x Where the slope = 6	$m = 3x^2$	
	$6 = 3x^2$	
	$x^2 = 2$	
	$x = \pm\sqrt{2}$	
Now, find $f(x)$	$f(x) = x^3$	
	$f(\sqrt{2}) = (\sqrt{2})^3 = 2^{\frac{3}{2}} \approx 2.83$	
	$f(-\sqrt{2}) = (-\sqrt{2})^3 \approx -2.83$	
Two Points	$(x, y) = (\sqrt{2},\ 2.83)$	
	$(x, y) = (-\sqrt{2}, -2.83)$	

Trig Topics – Part 1

Right Triangles

Right Triangles
Right Triangles have one angle = 90°

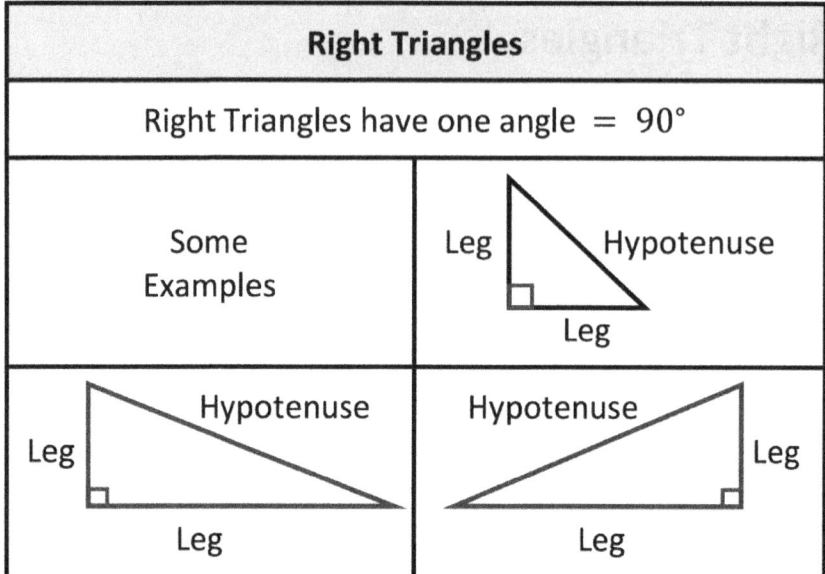

Some Examples	Leg / Hypotenuse / Leg
Leg / Hypotenuse / Leg	Hypotenuse / Leg / Leg

In a right triangle, the side opposite the right angle is the hypotenuse. The sides adjacent (next to) the right angle are often called the "legs."

The Hypotenuse is always the largest of the 3 sides.

Pythagorean Theorem – For Right Triangles

If the length of the hypotenuse, of a right triangle, is c and the lengths of the other two sides (legs) are a and b, then:

$$c^2 = a^2 + b^2$$

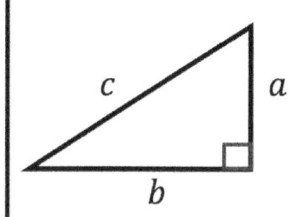

Right Triangles – Infinite Possibilities!

The other two angles, in a right triangle must add up to 90° because the sum of the angles, in any triangle, is $180°$. There are infinite possibilities with the measure of the other two angles. For example:

90, 45, 45   or   90, 20, 70   or   90, 19.2, 70.8 ...

There are three right triangles that are used often. So, it's a good idea to be familiar with them.

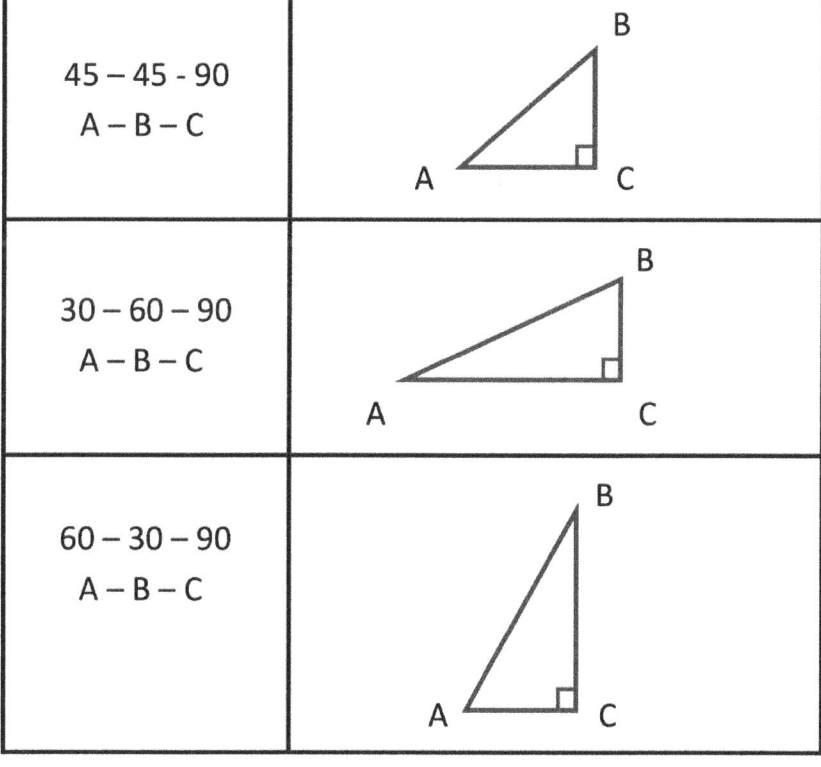

45 – 45 - 90
A – B – C

30 – 60 – 90
A – B – C

60 – 30 – 90
A – B – C

3 "Popular" Right Triangles, Hypotenuse = 1

If the hypotenuse = 1 then the other two lengths are:

45 – 45 - 90 A – B – C	Hypotenuse (AB) = 1, BC = $\frac{\sqrt{2}}{2}$, AC = $\frac{\sqrt{2}}{2}$
30 – 60 – 90 A – B – C	Hypotenuse (AB) = 1, BC = $\frac{1}{2}$, AC = $\frac{\sqrt{3}}{2}$
60 – 30 – 90 A – B – C	Hypotenuse (AB) = 1, BC = $\frac{\sqrt{3}}{2}$, AC = $\frac{1}{2}$

Right Triangles – 3 "Popular" Triangles	
If the hypotenuse $= h$ then the other two lengths are:	
45 – 45 - 90 A – B – C	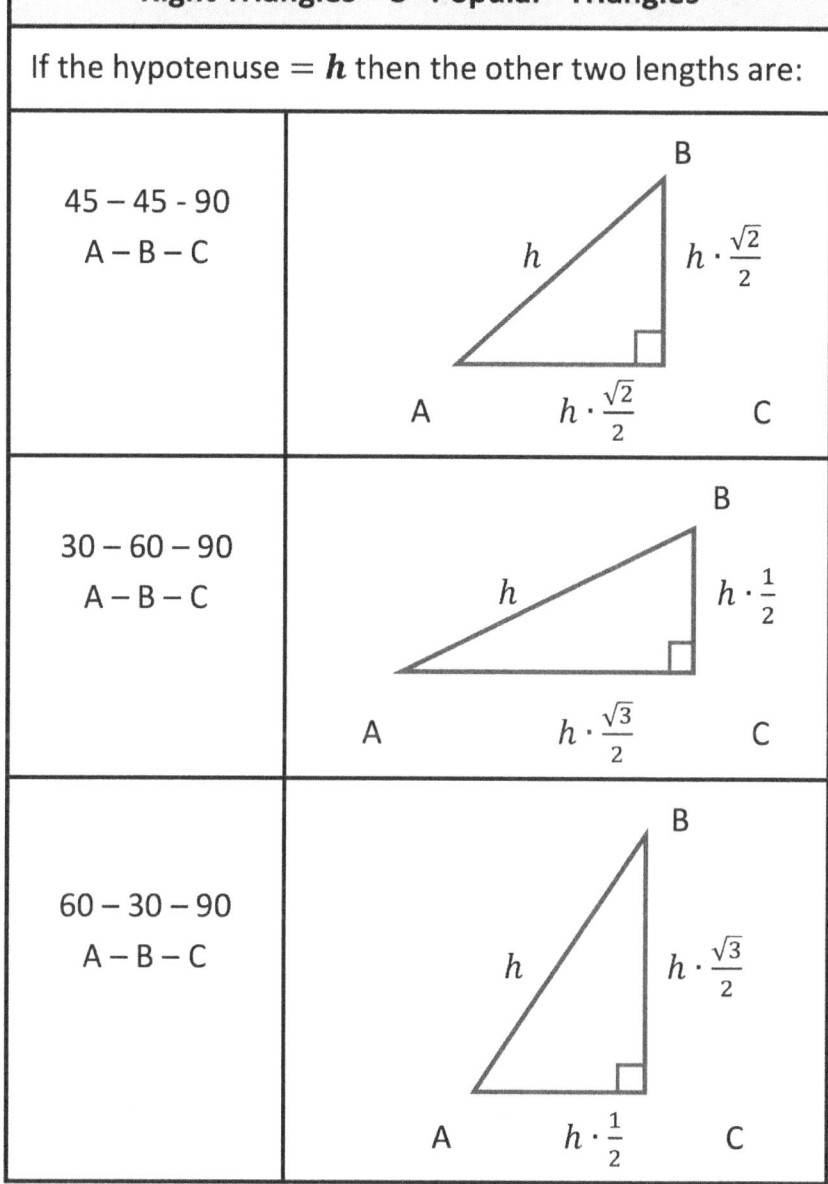
30 – 60 – 90 A – B – C	
60 – 30 – 90 A – B – C	

3 "Popular" Right Triangles, Hypotenuse = 1 Confirm – Ex. 1	
Use the Pythagorean Theorem to confirm the lengths.	
45 – 45 – 90	
$1^2 = \left(\frac{\sqrt{2}}{2}\right)^2 + \left(\frac{\sqrt{2}}{2}\right)^2$ $1 = \frac{2}{4} + \frac{2}{4}$ TRUE	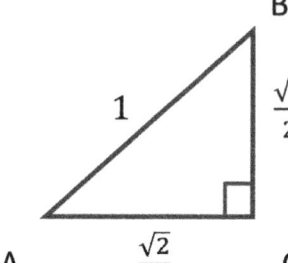
30 – 60 – 90	
$1^2 = \left(\frac{1}{2}\right)^2 + \left(\frac{\sqrt{3}}{2}\right)^2$ $1 = \frac{1}{4} + \frac{3}{4}$ TRUE	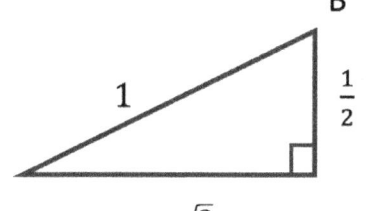
60 – 30 – 90	
$1^2 = \left(\frac{\sqrt{3}}{2}\right)^2 + \left(\frac{1}{2}\right)^2$ $1 = \frac{3}{4} + \frac{1}{4}$ $1 = 1$ TRUE	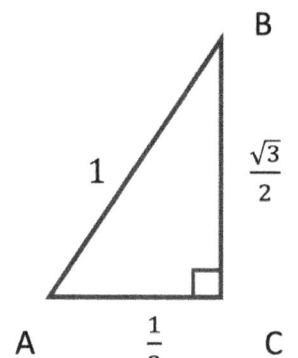

Right Triangles
Use Pythagorean Theorem – Ex. 2

Given a right triangle with leg lengths of 3 and 4
Use the Pythagorean Theorem to find the hypotenuse.

Make a sketch	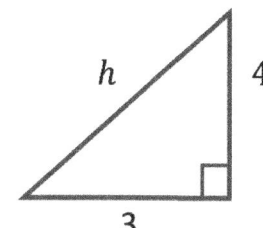
Use Pythagorean Theorem $$c^2 = a^2 + b^2$$	$h^2 = 3^2 + 4^2$ $h^2 = 9 + 16$ $h^2 = 25$ $h = \pm\sqrt{25}$ $h = 5$ Answer

Note #1: Disregard the negative solution.

Note #2: The 3-4-5 triangle is also very "Popular." It appears in many standardized exams. It may appear as a 3-4-5 triangle, or multiples of it (which are similar triangles). For example: 6-8-10 or 9-12-15 ...

The Unit Circle

Right Triangles in a Circle
With Radius = 1

There are an infinite number of triangles inside any circle. The diagram below shows 3 "Popular" or standard triangles within a circle, with radius = 1.

The 3 popular triangles are:
    45-45-90,   30-60-90,   and   60-30-90

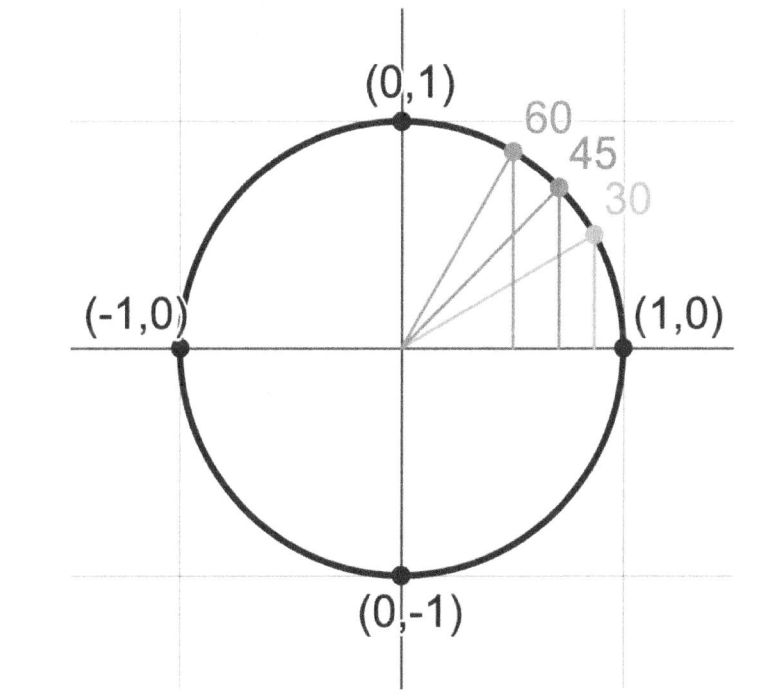

Can you find the coordinates where the 3 triangles intersect the circle? (See next page!)

Right Triangles in a Circle
With Radius = 1   (Showing Coordinates)

There are an infinite number of triangles inside any circle. The diagram below shows 3 standard triangles within a circle, with radius = 1.

The 3 popular triangles are:
    45-45-90,   30-60-90,   and   60-30-90

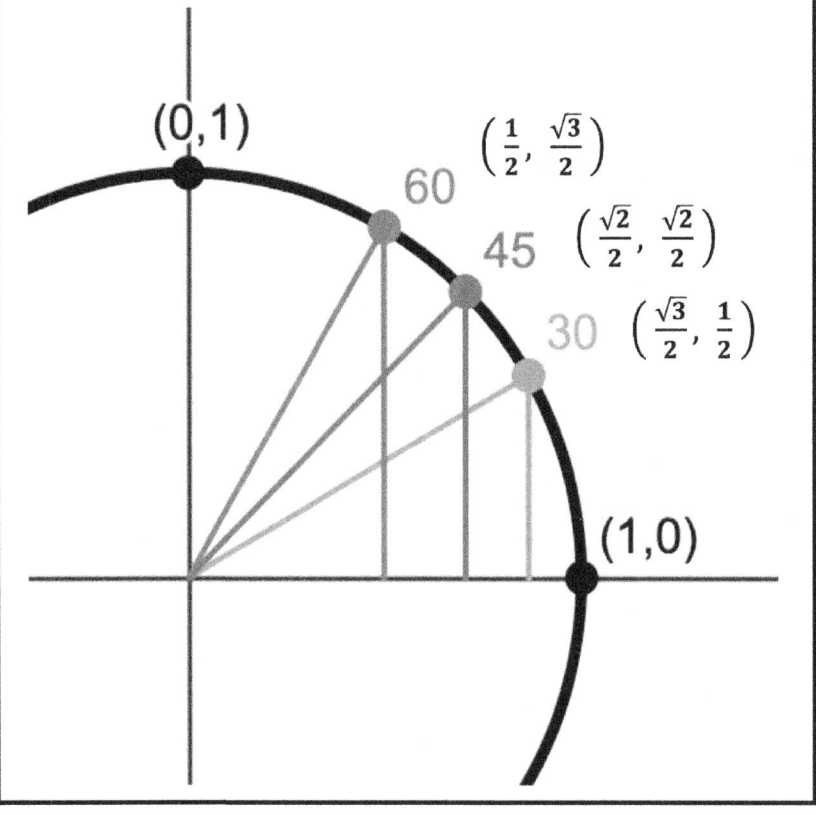

The Unit Circle
With Radius = 1 (Quadrant 1)

There are an infinite number of triangles inside any circle. The diagram below shows 3 standard triangles within a circle, with radius = 1.

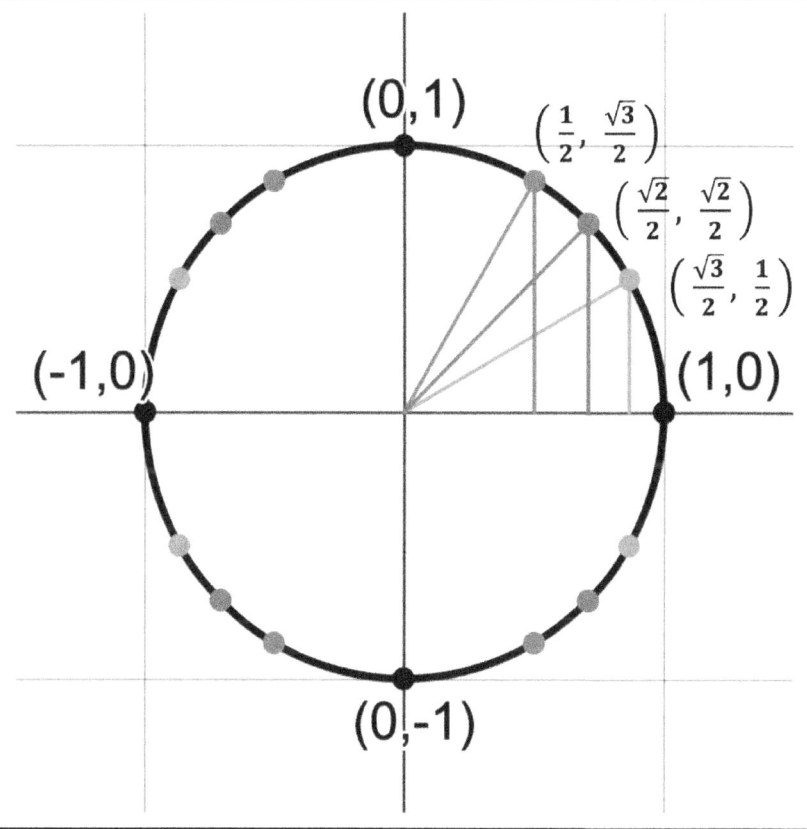

Can you find the coordinates of all points on the unit circle? (See next page.)

The Unit Circle
With Radius = 1

There are an infinite number of triangles inside any circle. The diagram below shows 3 right triangles $(30°, 45°, 60°)$ within a circle, with radius = 1.

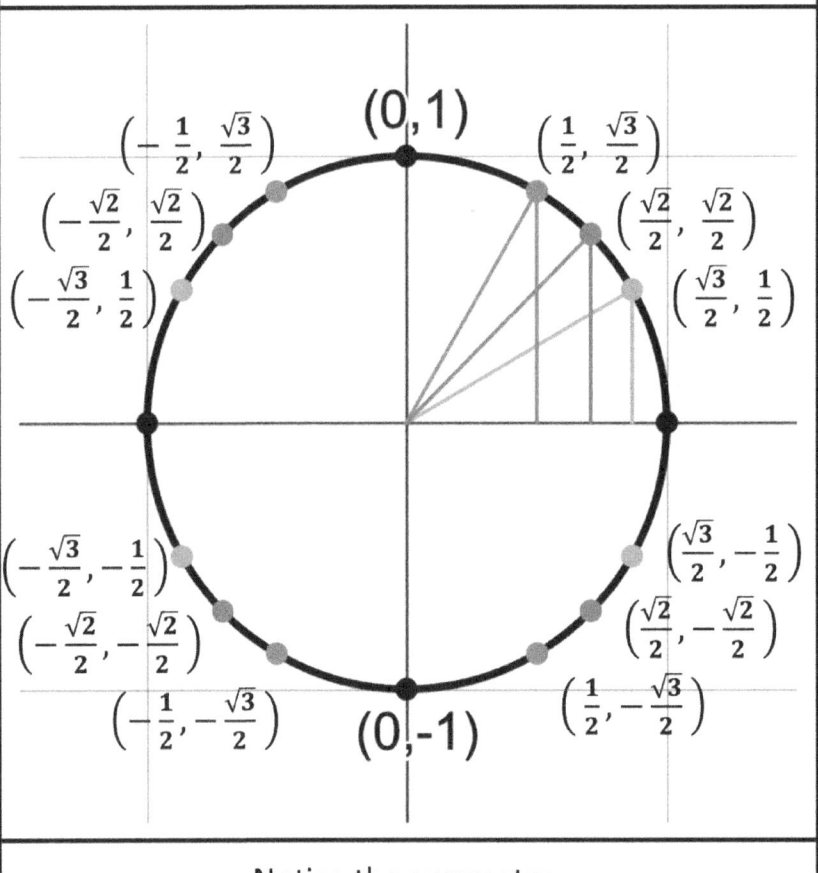

Notice the symmetry.

The Unit Circle – Measuring Angles

Angles, within a circle, are measured from the x axis, as shown below.

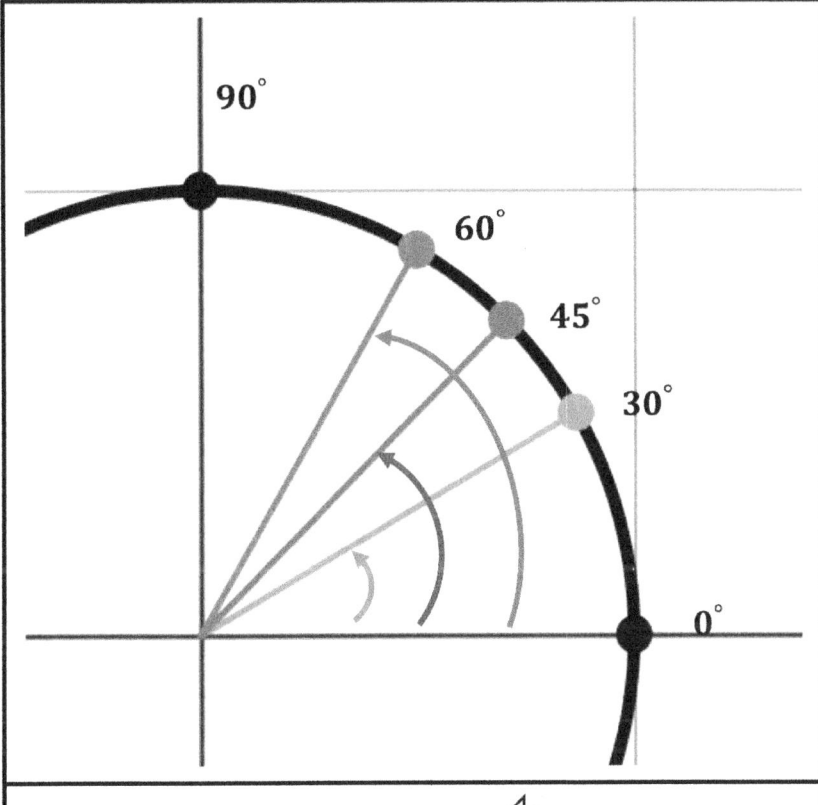

Like opening a book!

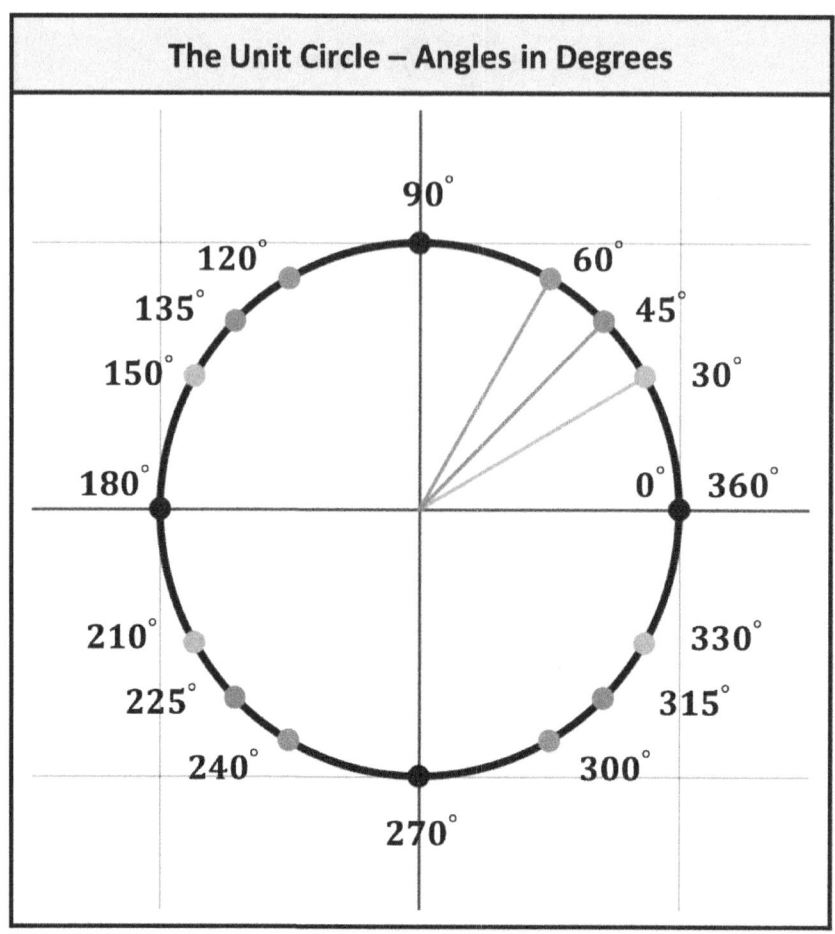

The Unit Circle – Angles in Radians

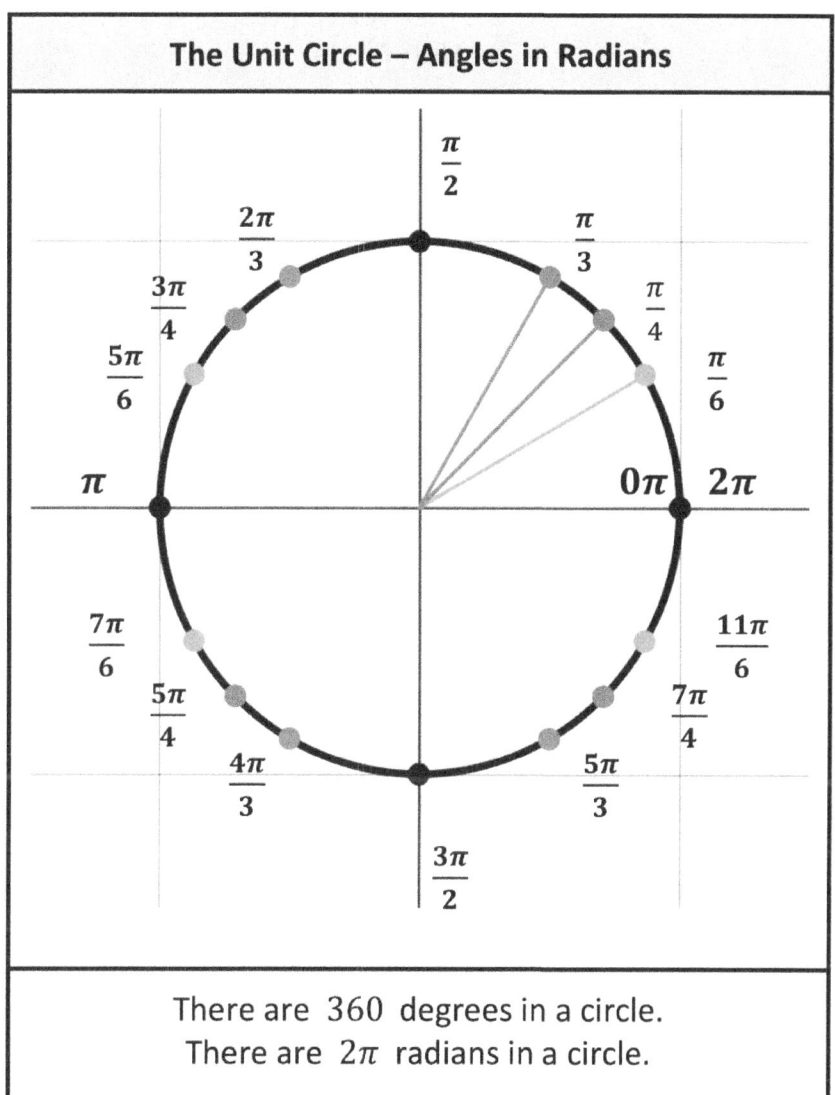

There are 360 degrees in a circle.
There are 2π radians in a circle.

$360° = 2\pi$ radians.
$180° = \pi$ radians.

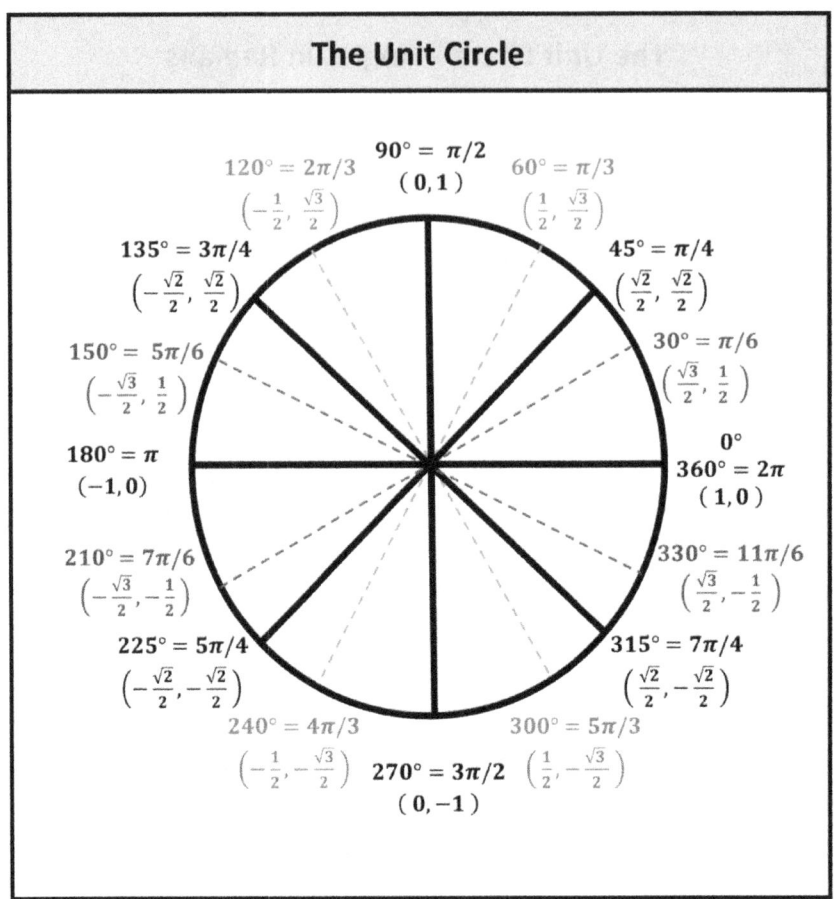

Trig. Function	When $r = 1$
$x = r \cos \theta$	$x = \cos \theta$
$y = r \sin \theta$	$y = r \sin \theta$

Trigonometric Functions

Trigonometric Functions

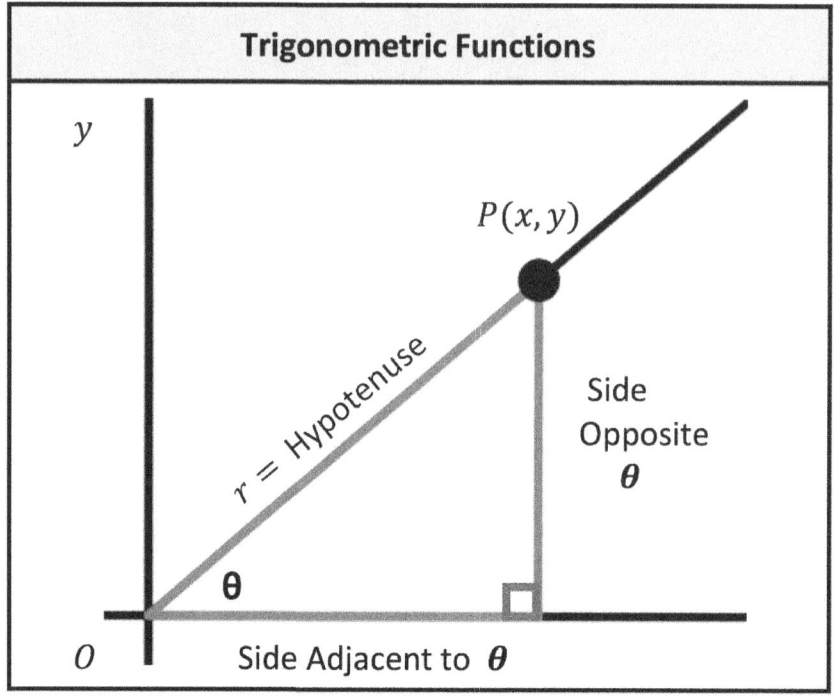

6 Trigonometric Functions

$\cos\theta = \dfrac{Adjacent}{r} = \dfrac{x}{r}$	$\sec\theta = \dfrac{1}{\cos\theta} = \dfrac{r}{x}$
$\sin\theta = \dfrac{Opposite}{r} = \dfrac{y}{r}$	$\csc\theta = \dfrac{1}{\sin\theta} = \dfrac{r}{y}$
$\tan\theta = \dfrac{Opposite}{Adjacent} = \dfrac{y}{x}$	$\cot\theta = \dfrac{1}{\tan\theta} = \dfrac{x}{y}$

$x = r\cos\theta$	$y = r\sin\theta$

Inverse Trigonometric Functions

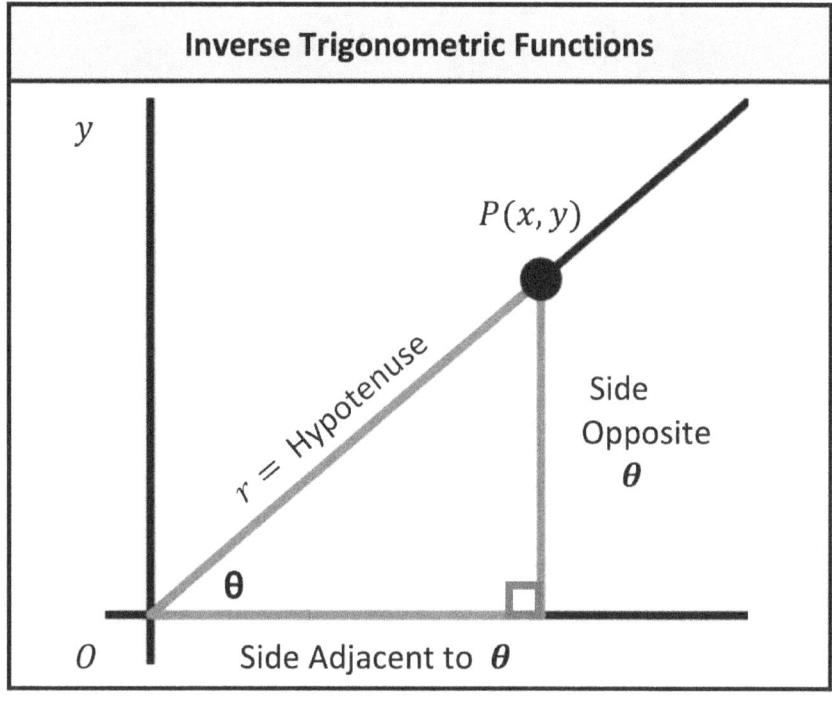

Inverse Trig. Functions. (2 notations)
$\theta = \cos^{-1}\left(\dfrac{x}{r}\right) = arcCos\left(\dfrac{x}{r}\right)$
$\theta = \sin^{-1}\left(\dfrac{y}{r}\right) = arcsin\left(\dfrac{y}{r}\right)$
$\theta = \tan^{-1}\left(\dfrac{y}{x}\right) = arctan\left(\dfrac{y}{x}\right)$
$\theta = \cos^{-1}\left(\dfrac{x}{r}\right)$ = "The angle whose cosine is $\left(\dfrac{x}{r}\right)$"

Trigonometric Functions – Ex. 1

Given: An acute angle θ in standard position and its terminal side passes through $P(4,5)$.
Find: $\cos\theta, \sin\theta,$ and $\tan\theta$.

Notes:
- Acute means the angle is $< 90°$
- Obtuse means the angle is $> 90°$
- Standard position means the vertex is at the orgin.

Make a sketch and use Pythagorean Theorem to find the Hypotenuse	(sketch: right triangle with legs 4 and 5, hypotenuse r, angle θ)	$r^2 = 4^2 + 5^2$ $r^2 = 16 + 25$ $r = \pm\sqrt{41}$ $r = \sqrt{41}$
$\cos\theta$	$\cos\theta = \frac{x}{r} = \frac{4}{\sqrt{41}}\left(\frac{\sqrt{41}}{\sqrt{41}}\right) = \frac{4\sqrt{41}}{41}$	
$\sin\theta$	$\sin\theta = \frac{y}{r} = \frac{5}{\sqrt{41}}\left(\frac{\sqrt{41}}{\sqrt{41}}\right) = \frac{5\sqrt{41}}{41}$	
$\tan\theta$	$\tan\theta = \frac{y}{x} = \frac{5}{4}$	

Trigonometric Functions – Ex. 2

Given: An acute angle θ in standard position and its terminal side passes through $P(5,12)$.
Find: The six trigonometric functions.

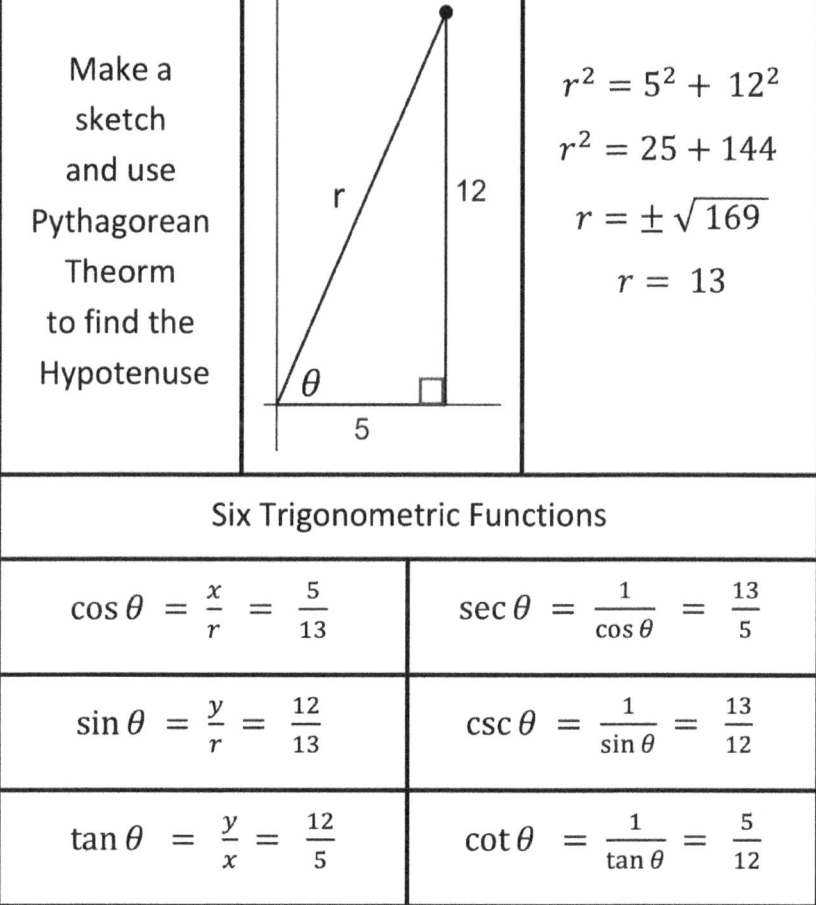

Make a sketch and use Pythagorean Theorm to find the Hypotenuse

$r^2 = 5^2 + 12^2$
$r^2 = 25 + 144$
$r = \pm\sqrt{169}$
$r = 13$

Six Trigonometric Functions

$\cos\theta = \dfrac{x}{r} = \dfrac{5}{13}$	$\sec\theta = \dfrac{1}{\cos\theta} = \dfrac{13}{5}$
$\sin\theta = \dfrac{y}{r} = \dfrac{12}{13}$	$\csc\theta = \dfrac{1}{\sin\theta} = \dfrac{13}{12}$
$\tan\theta = \dfrac{y}{x} = \dfrac{12}{5}$	$\cot\theta = \dfrac{1}{\tan\theta} = \dfrac{5}{12}$

Trigonometric Functions – Ex. 3

Given: $\sin\theta = \frac{1}{3}$ and θ is an acute angle.

Find: $\cos\theta$ and $\tan\theta$

If $\sin\theta = \frac{1}{3}$

Then $\frac{y}{r} = \frac{1}{3}$

So:
$y = 1$ and $r = 3$

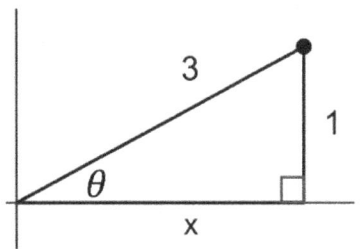

Use Pythagorean Theorem to find the missing side.

$r^2 = x^2 + y^2$

$3^2 = x^2 + 1^2$

$x^2 = 3^2 - 1^2$

$x^2 = 9 - 1$

$x = \pm\sqrt{8}$

$x = 2\sqrt{2}$

$$\cos\theta = \frac{x}{r} = \frac{2\sqrt{2}}{3}$$

$$\tan\theta = \frac{y}{x} = \frac{1}{2\sqrt{2}}\left(\frac{\sqrt{2}}{\sqrt{2}}\right) = \frac{\sqrt{2}}{4}$$

Trigonometric Functions – Ex. 4

Use information in the diagram to find the lengths of sides.

$\overline{BC}$ and $\overline{AB}$

Use a calculator

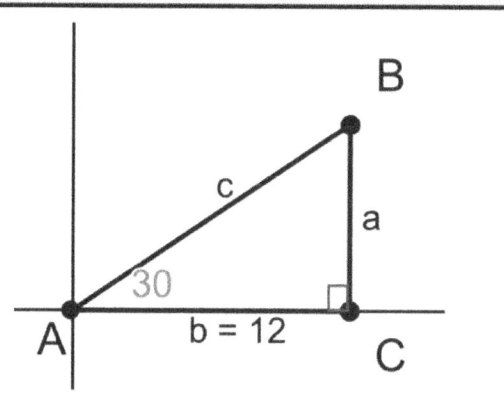

Find a Note: $a = \overline{BC}$	$\tan 30 = \dfrac{a}{12}$ $a = 12 \cdot \tan 30$ $a = 12 \cdot \left(\dfrac{\sqrt{3}}{3}\right) = 4\sqrt{3}$
Find c Note: $c = \overline{AB}$	$\cos 30 = \dfrac{b}{c}$ $c = \dfrac{b}{\cos 30}$ $c = \dfrac{12}{\left(\frac{\sqrt{3}}{2}\right)} = \dfrac{24}{\sqrt{3}}$ $c = \dfrac{24}{\sqrt{3}} \left(\dfrac{\sqrt{3}}{\sqrt{3}}\right) = \dfrac{24\sqrt{3}}{3} = 8\sqrt{3}$

Trigonometric Functions – Ex. 5a

Use information in the diagram to find the measure of angle A Use a calculator	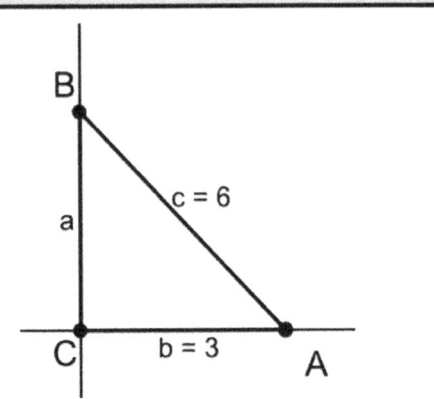

What do we know? Hint: Use given information.	For $\theta = A$ $\cos \theta = \dfrac{adjacent}{hypotenuse}$ $\cos A = \dfrac{b}{c}$ $\cos A = \dfrac{3}{6} = \dfrac{1}{2}$
We know angle A Is an angle whose cosine is $\dfrac{1}{2}$	$A = \cos^{-1}\left(\dfrac{1}{2}\right)$ $A = 60°$ Use calculator. Make sure calculator MODE is in Degrees.

Trigonometric Functions – Ex. 5b	
Use information in the diagram to find the measure of angle B Use a calculator	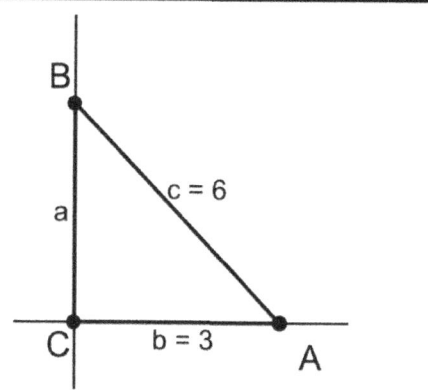

What do we know? Hint: Use given information.	For $\theta = B$ $\sin \theta = \dfrac{opposite}{hypotenuse}$ $\sin B = \dfrac{b}{c}$ $\sin B = \dfrac{3}{6} = \dfrac{1}{2}$
We know angle B Is an angle whose sine is $\dfrac{1}{2}$	$B = \sin^{-1}\left(\dfrac{1}{2}\right)$ $B = 30°$ Use calculator. Make sure calculator MODE is in Degrees.

Trig Functions of General Angles

Trigonometric Functions of General Angles

Place θ in standard position, choose a point $P(x, y)$ on the terminal side of θ, and let $r = $ the distance $\overline{OP}$.

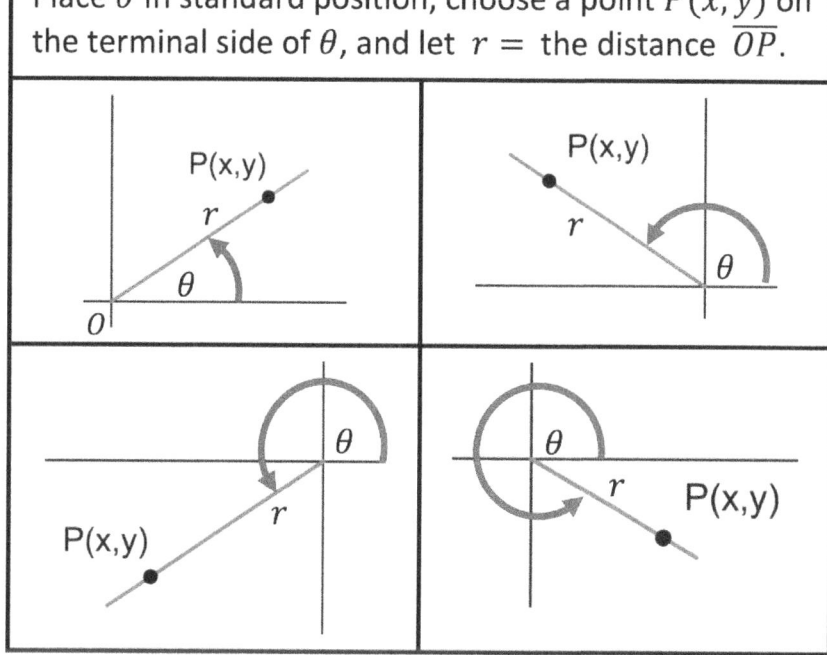

$\cos \theta = \frac{x}{r}$	$\sec \theta = \frac{1}{\cos \theta} = \frac{r}{x} \; ; \; x \neq 0$
$\sin \theta = \frac{y}{r}$	$\csc \theta = \frac{1}{\sin \theta} = \frac{r}{y} \; ; \; y \neq 0$
$\tan \theta = \frac{y}{x} \; ; \; x \neq 0$	$\cot \theta = \frac{1}{\tan \theta} = \frac{x}{y} \; ; \; y \neq 0$

$$r = \sqrt{x^2 + y^2}$$

Trigonometric Functions of General Angles
Reference Angle

When θ is in standard position, we call α the reference angle. Where α is an acute angle that represents θ. See diagrams, below.

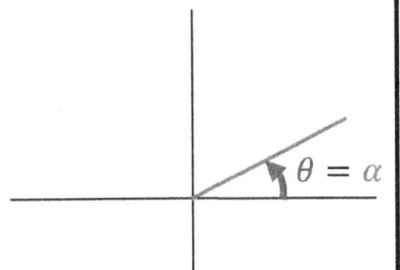

Q1: $\theta = 30\,;\, \alpha = 30$

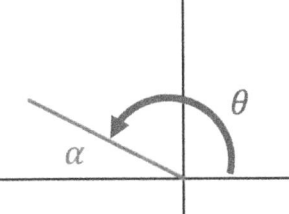

Q2: $\theta = 150\,;\, \alpha = 30$

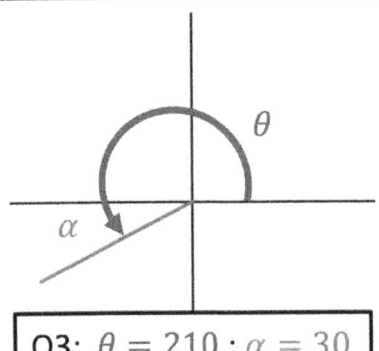

Q3: $\theta = 210\,;\, \alpha = 30$

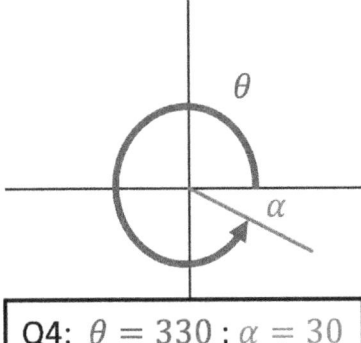

Q4: $\theta = 330\,;\, \alpha = 30$

Trigonometric Functions of General Angles
Reference Angles – Ex. 1

Find the quadrant and measure of the reference angles, α, for the following angles:

$\theta = 140°$		Q2 $\alpha = 180 - 140$ $\alpha = 40°$
$\theta = 300°$		Q4 $\alpha = 360 - 300$ $\alpha = 60°$
$\theta = -135°$		Q3 $\alpha = 180 - 135$ $\alpha = 45°$
$\theta = 480°$		Q2 $\theta = 480 - 360$ $\theta = 120$ $\alpha = 180 - 120$ $\alpha = 60°$

Trigonometric Functions of General Angles
Reference Angles – Ex. 2

Find the quadrant and exact value of the following:
$$\tan 330° \quad \text{and} \quad \csc(-225°)$$

$\theta = 330°$

Q4

$\alpha = 360 - 330$

$\alpha = 30°$

$\tan 30° = \dfrac{\sqrt{3}}{3} = \dfrac{y}{x}$

In Q4, $(x, y) = (+, -)$

$\tan 330° = -\dfrac{\sqrt{3}}{3}$ Answer

$\theta = -225°$

Q2

$\alpha = 225 - 180$

$\alpha = 45°$

$\csc 45 = \dfrac{1}{\sin 45} = \dfrac{2}{\sqrt{2}}\left(\dfrac{\sqrt{2}}{\sqrt{2}}\right) = \sqrt{2} = \dfrac{r}{y}$

In Q2, $(x, y) = (-, +)$

$\csc(-225°) = \sqrt{2}$ Answer

Trigonometric Functions of General Angles
Reference Angles – Ex. 3

Find the five other trigonometric functions of θ if ...

$$\cos\theta = -\frac{2}{5} \quad \text{and} \quad 180° < \theta < 360°$$

Negative cosine tells us θ is in Q2 or Q3.

$180° < \theta < 360°$ tells us θ is in Q3 or Q4.

Therefore, θ is in Q3.   In Q3, $(x, y) = (-, -)$

| Sketch a reference angle in Q1, based on $\cos\alpha = \frac{2}{5} = \frac{x}{r}$ | $x^2 + y^2 = r^2$ $2^2 + y^2 = 5^2$ $y = \pm\sqrt{5^2 - 2^2}$ $y = \sqrt{21}$ | 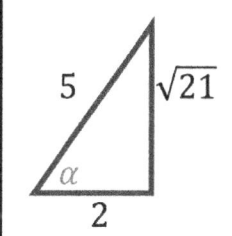 |

Remember: In Q3, $(x, y) = (-, -)$

$\cos\theta = \frac{x}{r} = -\frac{2}{5}$	$\sec\theta = \frac{1}{\cos\theta} = -\frac{5}{2}$
$\sin\theta = \frac{y}{r} = -\frac{\sqrt{21}}{5}$	$\csc\theta = \frac{1}{\sin\theta} = -\frac{5}{\sqrt{21}} = -\frac{5\sqrt{21}}{21}$
$\tan\theta = \frac{y}{x} = \frac{\sqrt{21}}{2}$	$\cot\theta = \frac{1}{\tan\theta} = \frac{2}{\sqrt{21}} = \frac{2\sqrt{21}}{21}$

Trigonometric Functions of General Angles – Ex. 4

A sailor sees a light tower, with the top of the tower at an angle of elevation of 25.6°. After his ship has moved 1050 m closer, the angle of elevation is 31.2°. What is the height of the light tower, above sea level?

Make a sketch	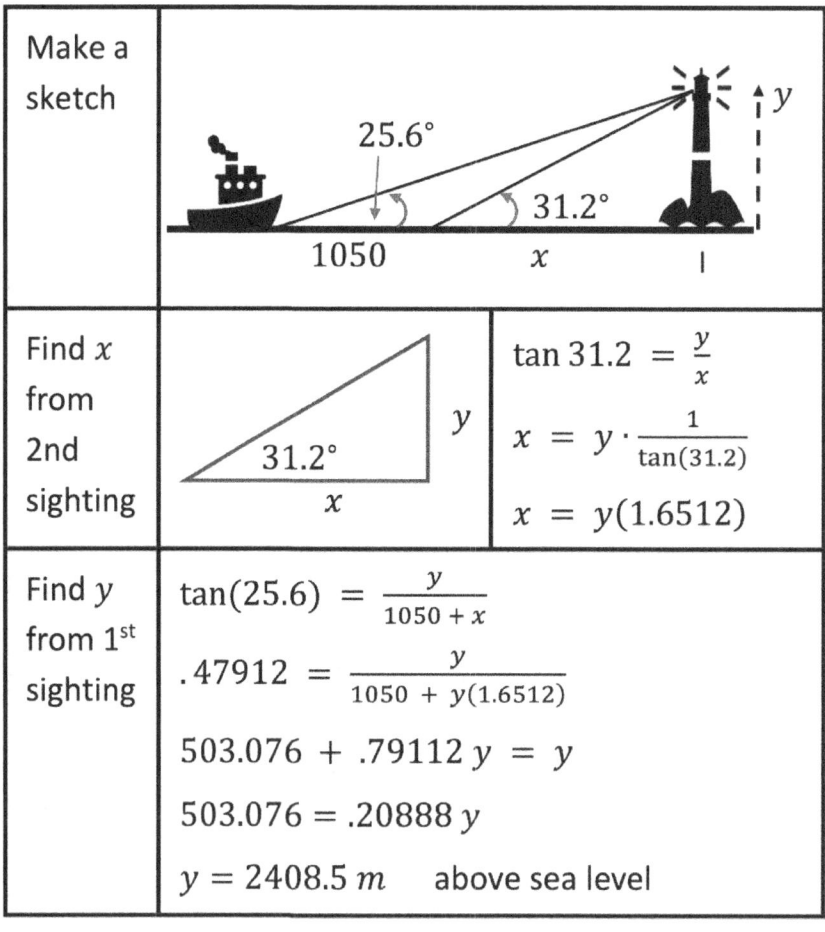
Find x from 2nd sighting	$\tan 31.2 = \dfrac{y}{x}$ $x = y \cdot \dfrac{1}{\tan(31.2)}$ $x = y(1.6512)$
Find y from 1st sighting	$\tan(25.6) = \dfrac{y}{1050 + x}$ $.47912 = \dfrac{y}{1050 + y(1.6512)}$ $503.076 + .79112\, y = y$ $503.076 = .20888\, y$ $y = 2408.5\ m$ above sea level

Trig Functions and Baseball – Ex. 5a

A baseball is hit at a point 3 feet above the ground at a velocity of 100 feet per second at an angle of $50°$. Write equations to model the vertical height (y) and the horizontal distance (x) of the ball as a function of time.   Use Gravity $= 32 \frac{ft}{s^2}$

Make a sketch

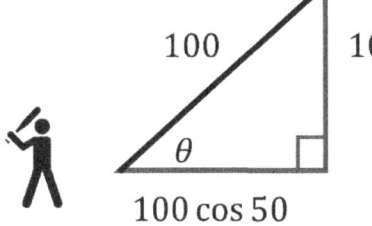

$y = -$ (gravity effect) + (vert. hit effect) + (init. height)

$y = -\left(\frac{32}{2}\right) t^2 + (100 \sin 50) t + (3)$

$x =$ (horizontal hit effect)

$x = (100 \cos 50) t$

Note the units	$(vel.)(time) = \left(\frac{ft}{s}\right) \cdot (s) = ft$
	$(accel.)(time^2) = \left(\frac{ft}{s^2}\right) \cdot (s^2) = ft$

	Trig Functions and Baseball – Ex. 5b
	A baseball is hit at a point 3 feet above the ground at a velocity of 100 feet per second at an angle of $50°$. There is a 5 foot fence, located 300 feet from home plate. Will the baseball clear the fence?
Previously found	$y = -(16)t^2 + (100 \sin 50)t + (3)$ $x = (100 \cos 50)t$
Find time when ball reaches fence	$300 = (100 \cos 50)t$ $t = \dfrac{300}{100 \cos 50} = 4.7$ seconds
Find height when ball reaches fence	$y = -(16)t^2 + (100 \sin 50)t + (3)$ $y = -(16)(4.7)^2 +$ $\qquad + (100 \sin 50)(4.7) + (3)$ $y = -353.4 + 360.0 + 3$ $y = 9.6$ ft.
Answer	Yes! The baseball will clear the 5 ft. fence by 4.6 feet.

Law of Cosines

Law of Cosines – For All Triangles
If you know the lengths of two sides and the measure of the included angle, then the **Law of Cosines** can be used to find the length of the other side. **SAS**
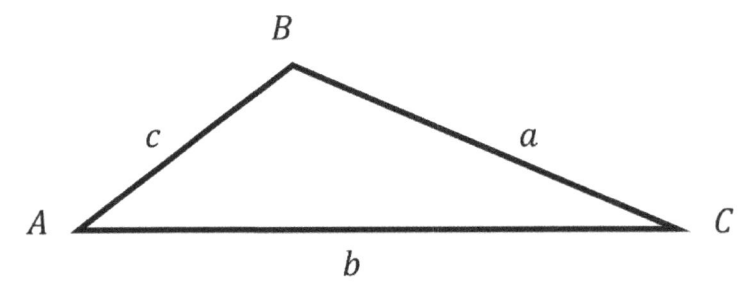

Note: The naming convention. Angles are marked with capitol letters. The sides, opposite from the angles are marked with lower-case letters.
The Law of Cosines is similar to the Pythagorean Theorem (for right triangles only). The Law of Cosines is for all triangles. The general form is: $$a^2 = b^2 + c^2 - correction$$
$$a^2 = b^2 + c^2 - 2bc \cdot \cos A$$
$$b^2 = a^2 + c^2 - 2ac \cdot \cos B$$
$$c^2 = a^2 + b^2 - 2ab \cdot \cos C$$

Law of Cosines – For All Triangles – Ex. 1

In $\triangle ABC$, $a = 10$, $b = 13$, $\angle C = 70°$
Find c to four significant digits.

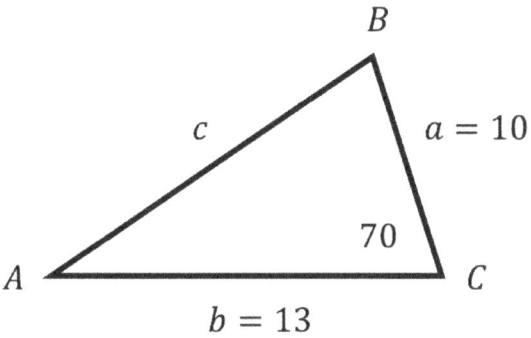

Note: We are looking for "c" so start with a version of the Law of Cosines with "C" on the ends (book-ends).

$c^2 = a^2 + b^2 - 2ab \cdot \cos C$

$c^2 = 10^2 + 13^2 - 2(10)(13) \cdot \cos(70)$

$c^2 = 100 + 169 - 260 \cdot \cos(70)$

$c^2 = 269 - 260 \cdot (.34202)$

$c^2 = 269 - 88.9252$

$c = \pm\sqrt{180.075} = 13.4192$

$c = 13.42$ to four significant digits.

Law of Cosines – All Sides Known – Ex. 2

A triangular-shaped lot has three sides with lengths: 150, 120, *and* 50 *meters*. Find the largest angle.

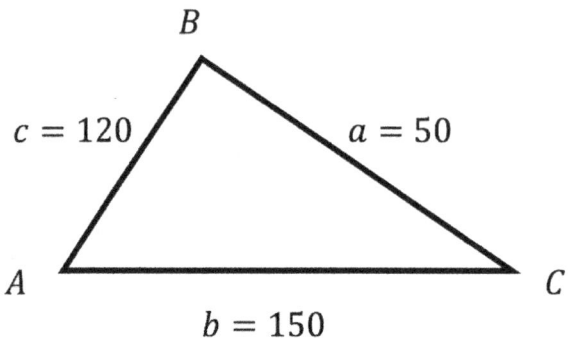

In our diagram, b is the longest side. So, angle B will be the largest angle. We are looking for "B" so start with a version of the Law of Cosines with "B" on the ends (book-ends).

$b^2 = a^2 + c^2 - 2ac \cdot \cos B$

$150^2 = 50^2 + 120^2 - 2(50)(120) \cdot \cos B$

$22500 = 16900 - 12000 \cdot \cos B$

$\cos B = \dfrac{22500 - 16900}{-12000} = \dfrac{5600}{-12000} = -.46667$

$B = \cos^{-1}(-.46667) = 117.82°$

Law of Sines

Law of Sines – For All Triangles

If you know the lengths of two sides and the measure of one angle, **NOT** between the two known sides, then the **Law of Sines** can be used to find the length of the other side. **SSA (or ASS)**

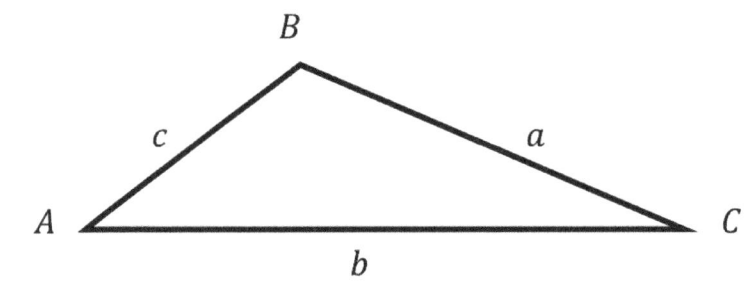

Note: The Law of Sines says the angles and the opposite sides are proportional.

$$\frac{a}{\sin A} = \frac{b}{\sin B} = \frac{c}{\sin C}$$

$$\frac{\sin A}{a} = \frac{\sin B}{b} = \frac{\sin C}{c}$$

Law of Sines – For All Triangles – Ex. 1a

In $\triangle ABC$ $\angle A = 40°$ and $a = 15$
Find $\angle B$ if $b = 20$

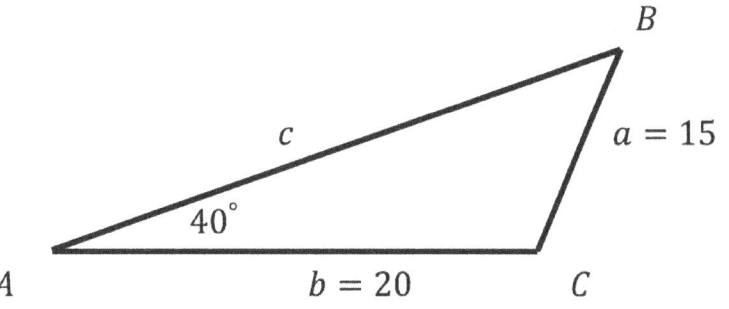

Note: We are looking for B so use a version of the Law of Sines with $\sin B$ in numerator of first term. Also, since we know $\sin A$ and a, use that ratio.

$$\frac{\sin B}{b} = \frac{\sin A}{a}$$

$$\frac{\sin B}{20} = \frac{\sin(40)}{15}$$

$$\sin B = 20\left(\frac{\sin(40)}{15}\right) = \frac{4}{3}\sin(40) = .85705$$

$$B = \sin^{-1}(.85705) = 59°$$

Law of Sines – For All Triangles – Ex. 1b

In $\triangle ABC$ $\angle A = 40°$ and $a = 15$
Find $\angle B$ if $b = 20$

Previously, we found
$\sin B = .85705$

But, there are two angles that have that $\sin$

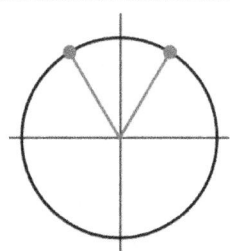

The calculator will return one value for
$B = \sin^{-1}(.85705) = 59°$
The default domain for the sin functions Is Q1 & Q4

There is another angle, in Q2, with the same sin
In Q2, the other angle is: $180 - 59 = 121°$
Therefore: $B = 59°$ or $121°$

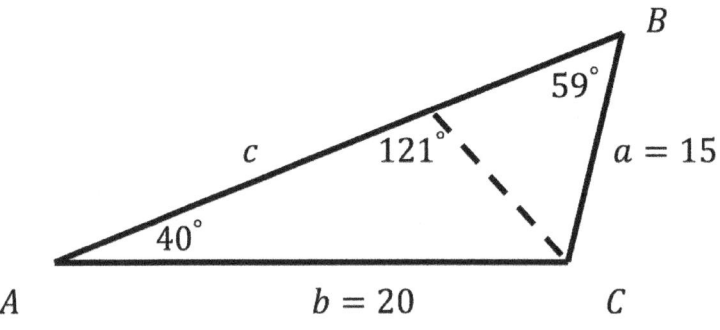

Law of Sines – For All Triangles – Ex. 2

A $123 \, ft$ support wire for a large flag pole makes an angle for $61°$ with the ground. This wire will be replaced with a new wire with an angle of $46°$ What will be the length of the new wire?

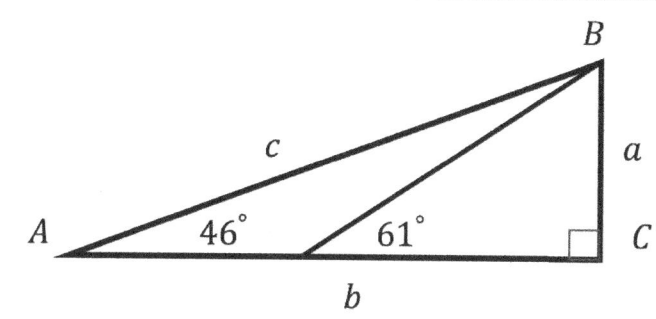

Use this eqn. For both cases.	$\dfrac{c}{\sin C} = \dfrac{a}{\sin A}$
Case 1: $A = 61°$ We know $c = 123$ Find a (pole height)	$\dfrac{123}{\sin 90} = \dfrac{a}{\sin 61}$ $a = \sin 61 \left(\dfrac{123}{\sin 90}\right) = 107.6$
Case 2: $A = 46°$ We know $a = 107.6$	$\dfrac{c}{\sin 90} = \dfrac{107.6}{\sin 46}$ $c = (1)\left(\dfrac{107.6}{\sin 46}\right) = 149.6 \, ft.$

General Triangles

General Triangles

In this section, examples will be provided to demonstrate working with General Triangles when given the following information:

Case	Given	Notes
SSS	Side-Side-Side	Fixed
SAS	Side-Angle-Side	Fixed
ASA, AAS	2 Angles, 1 Side	Fixed
SSA	Side-Side-Angle	Ambiguous

For all "Fixed" triangles, the approach is simple. If you are given a side and opposite angle, then use the "Law of Sines" to find additional information. The "Law of Sines" is a simpler equation to use than the "Law of Cosines."

For "Ambiguous" (or indeterminate) trinagles, be aware there are multiple possibilities. Visually place a hindge between the two given sides. There may be multiple possibilities for the angle between those two sides.

Default Domains for Trigonometric Functions

It is important to be aware of the default domains when using calculators with trigonometric functions. The default domains are listed below.

Trig Function	Default Domain
$\cos\theta$ and $\cos^{-1}(n)$	Q1 & Q2
$\sin\theta$ and $\sin^{-1}(n)$	Q1 & Q4
$\tan\theta$ and $\tan^{-1}(n)$	Q1 & Q4

When using inverse trig functions, there are usually 2 answers within 360° (infinite solutions for all angles). A calculator will return only one solution, within the default domain. Use your underatanding of trig functions to determine the additional solution.

	General Triangles (SSS) – Ex. 1
Solve $\triangle ABC$, Given: $a = 4$, $b = 6$, $c = 5$	
Make a sketch	Triangle with vertices A (bottom left), B (top), C (bottom right); $c = 5$, $a = 4$, $b = 6$
Can we use Law of Sines?	No! No angle and side pair. ☹
Use Law of Cosines to get more info.	$a^2 = b^2 + c^2 - 2bc \cdot \cos A$ $\cos A = \dfrac{a^2 - b^2 - c^2}{-2bc}$ $A = \cos^{-1}\left(\dfrac{a^2 - b^2 - c^2}{-2bc}\right) = 41.4°$
Now, use Law of Sines for more info.	$\dfrac{\sin B}{6} = \dfrac{\sin(41.4)}{4}$ $B = \sin^{-1}\left(6 \cdot \dfrac{\sin(41.1)}{4}\right) = 82.7°$
The last angle.	$C = 180 - 41.4 - 82.7 = 55.9°$

	General Triangles (SAS) – Ex. 2
Solve $\triangle ABC$, Given: $a = 8$, $c = 7$, $\angle B = 31.8°$	
Make a sketch	*Triangle with vertex B at top (angle 31.8), vertex A at bottom-left with side c = 7, vertex C at bottom-right with side a = 8, and side b between A and C.*
Can we use Law of Sines?	No! No angle and side pair. ☹
Use Law of Cosines to get more info.	$b^2 = a^2 + c^2 - 2ac \cdot \cos B$ $b^2 = 17.812$ $b = \sqrt{17.812} = 4.22$
Now, use Law of Sines for more info.	$\dfrac{\sin A}{8} = \dfrac{\sin(31.8)}{4.22}$ $A = \sin^{-1}\left(8 \cdot \dfrac{\sin(31.8)}{4.22}\right) = 87.4°$
The last angle.	$C = 180 - 31.8 - 87.4 = 60.8°$

General Triangles (AAS) – Ex. 3

Solve $\triangle ABC$, Given: $a = 40$, $\angle A = 45°$, $\angle C = 55°$

Make a sketch	*(triangle with vertices A (bottom-left, 45°), B (top), C (bottom-right, 55°); side $a = 40$ opposite A, side b along bottom, side c on left)*
Get 3rd angle	$B = 180 - 45 - 55 = 80°$
Can we use Law of Sines?	Yes! We have angle and side pair. ☺
Use Law of Sines to get more info.	$\dfrac{c}{\sin(55)} = \dfrac{40}{\sin(45)}$ $c = \sin 55 \left(\dfrac{40}{\sin 45}\right) = 46.3$
Use Law of Sines again.	$\dfrac{b}{\sin(80)} = \dfrac{40}{\sin(45)}$ $b = \sin 80 \left(\dfrac{40}{\sin 45}\right) = 55.7$

	General Triangles (ASS) – Ex. 4a
	Solve $\triangle ABC$, Given: $b = 22$, $c = 30$, $\angle B = 30°$
Make a sketch	*[Sketch of triangle ABC with vertex B at top right showing angle 30°, side $c = 30$ from A to B, side a from B to C, and base $b = 22$ from A to C]*
WARNING!! "ASS"	AMBIGUOUS CASE!!! The angle, between the two given sides (Angle A) can be several angles. Threat it like a hindge.
Consider multiple possiblities	*[Sketch showing triangle with hinge at A, dashed line indicating alternate position, $c = 30$, a, $b = 22$, angle $30°$ at B]*

General Triangles (ASS) – Ex. 4b		
Solve $\triangle ABC$, Given: $b = 22$, $c = 30$, $\angle B = 30°$		
Consider multiple possiblities	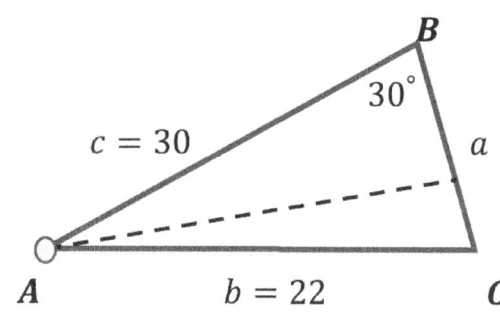	
Find C Because we know $c = 30$	$\dfrac{\sin C}{30} = \dfrac{\sin 30}{22}$ $\sin C = 30\left(\dfrac{\sin 30}{22}\right) = .681818$ $C = \sin^{-1}(.681818) = 43°$	
Find another value for angle C	There are two angles that have $\sin = .681818$	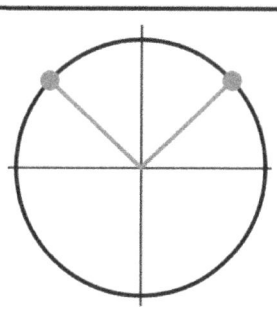
	Also: $C = 180 - 43 = 137°$	

General Triangles (ASS) – Ex. 4c
Solve $\triangle ABC$, Given: $b = 22$, $c = 30$, $\angle B = 30°$

Consider multiple possiblities	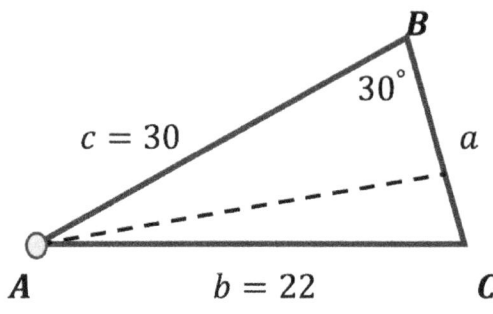
Previously found	$C = 43°$ or $C = 137°$

Case #1 $C = 43°$	$A = 180 - 30 - 43 = 107°$
	$\dfrac{a}{\sin 107} = \dfrac{22}{\sin 30}$
	$a = \sin 107 \left(\dfrac{22}{\sin 30}\right) = 42$

Case #2 $C = 137°$	$A = 180 - 30 - 137 = 13°$
	$\dfrac{a}{\sin 13} = \dfrac{22}{\sin 30}$
	$a = \sin 13 \left(\dfrac{22}{\sin 30}\right) = 9.9$

	General Triangles (ASS) – Ex. 5a
For: $\triangle ABC$, Given: $\angle A = 25°$, $c = 30$, $a = k$ Find values for k so that 0, 1, 2 triangles are possible.	
Make a sketch	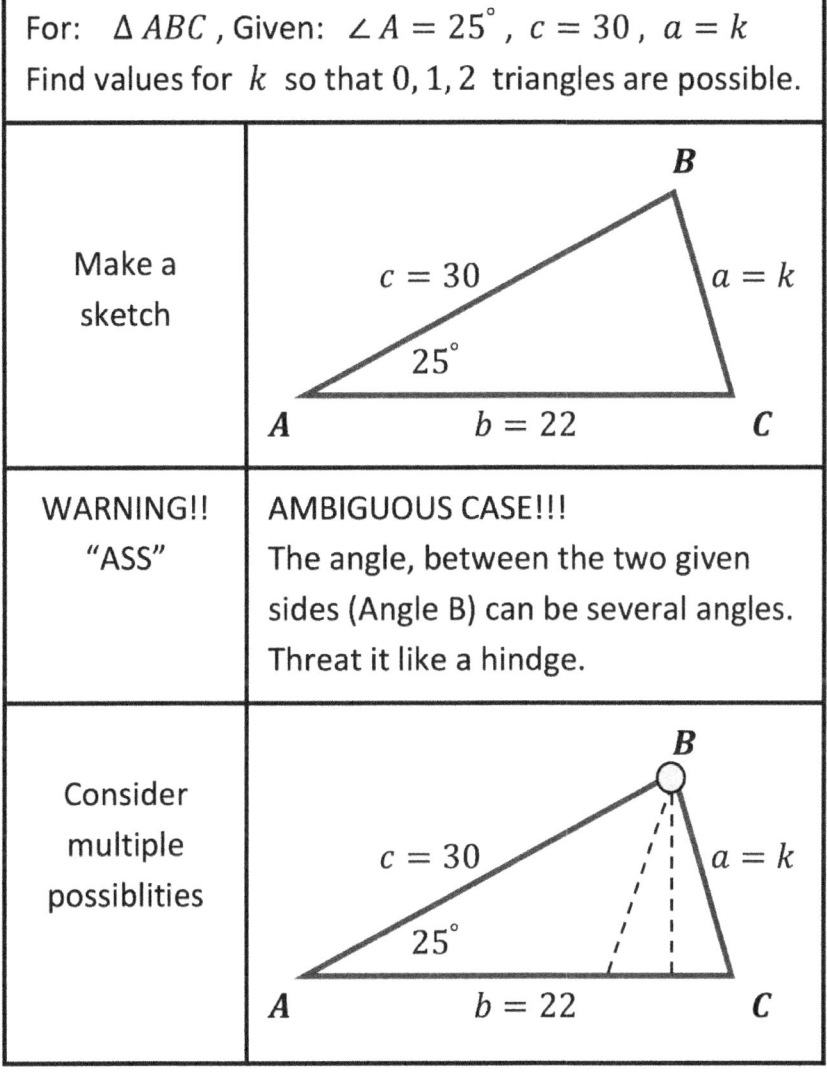
WARNING!! "ASS"	AMBIGUOUS CASE!!! The angle, between the two given sides (Angle B) can be several angles. Threat it like a hindge.
Consider multiple possiblities	

General Triangles (ASS) – Ex. 5b
For: $\triangle ABC$, Given: $\angle A = 25°$, $c = 30$, $a = k$ Find values for k so that 0, 1, 2 triangles are possible.

Consider multiple possiblities	(diagram: triangle with $c = 30$, angle $25°$ at A, $b = 22$ along base from A to C, and $a = k$ from B swinging to meet base)
0 Triangles Possible	If side a is too short, then no triangle is possible. Side a must be long enough to make a right angle at C.
	If $C = 90°$ (right triangle) $a = r \cdot \sin\theta$ $a = 30 \cdot \sin 25 = 12.7$
	0 Triangles if: $k < 12.7$
2 Triangles Possible	If: $12.7 < a < 30$ then side a could swing inward or outward to create two different triangles.
	2 Triangles if: $12.7 < k < 30$

	General Triangles (ASS) – Ex. 5c
For: $\triangle ABC$, Given: $\angle A = 25°$, $c = 30$, $a = k$ Find values for k so that $0, 1, 2$ triangles are possible.	
Consider multiple possiblities	*[Figure: Triangle with vertex A at lower left (angle $25°$), B at top, C at lower right. Side $c = 30$ from A to B, side $a = k$ from B to C, side $b = 22$ from A to C. Dashed lines show alternate positions of side a swinging from B.]*
Previously Found	0 Triangles if: $k < 12.7$
	2 Triangles if: $12.7 < k < 30$
1 Triangle Possible	If side $a = 12.7$ then $C = 90°$ and only one triangle would be possible.
	If side $a \geq 30$, it could not rotate inward. It could only rotate outward and only one triangle would be possible.
	1 Triangle if: $k = 12.7$ or $k \geq 30$

Bearings

Bearings

In navigation, directions are usually given in terms of bearings. Some examples are given, below.

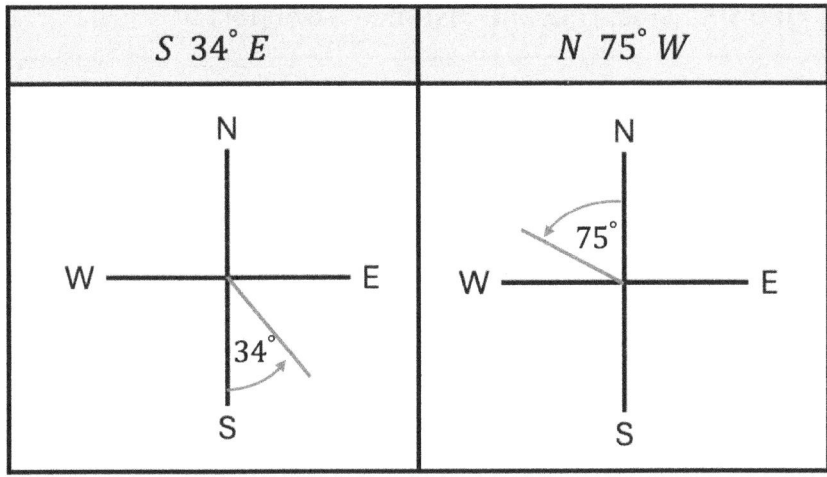

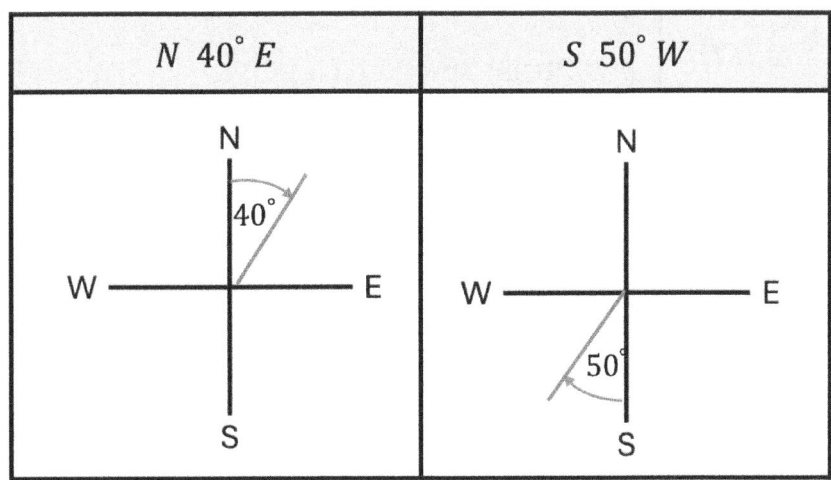

Bearings – Ex. 1a

A ship leaves port at noon and heads due west at 20 knots (nautical mph). At 2 PM the ship changes course to $N\ 54°\ W$ as shown in the diagram. Find ship's bearing and distance from port at 3 PM.

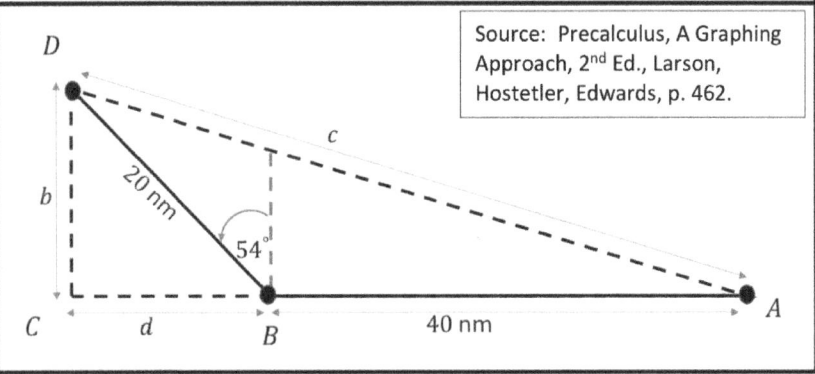

Source: Precalculus, A Graphing Approach, 2nd Ed., Larson, Hostetler, Edwards, p. 462.

$\triangle BCD$	$B = 90 - 54 = 36°$ $b = 20 \sin 36 = 11.76$ $d = 20 \cos 36 = 16.18$
$\triangle ACD$	$\tan A = \dfrac{b}{d+40} = \dfrac{11.76}{56.18} = .2093$ $A = \tan^{-1}(.2093) = 11.82°$ $D = 90 - 11.82 = 78.18°$ $\dfrac{c}{\sin 90} = \dfrac{11.76}{\sin 11.82} \rightarrow c = 57.9\ nm$

Bearings – Ex. 1b

A ship leaves port at noon and heads due west at 20 knots (nautical mph). At 2 PM the ship changes course to $N\ 54°\ W$ as shown in the diagram. Find ship's bearing and distance from port at 3 PM.

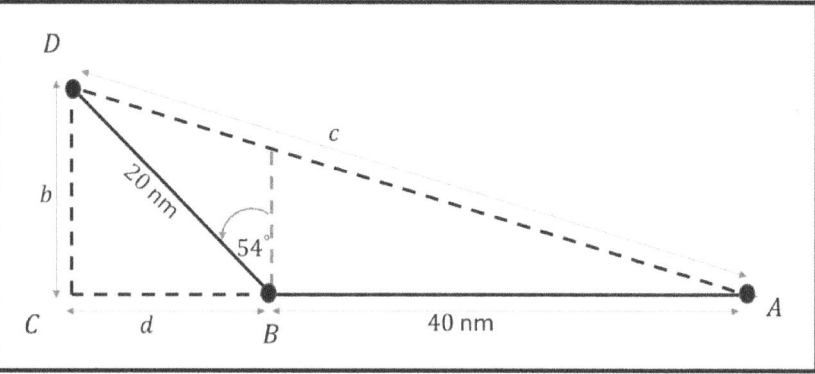

Previously Found In $\triangle ACD$	$A = 11.82°$ $D = 78.18°$	$c = 57.9$ nm
Distance from port	Distance $= 57.9$ nm	
Bearings from port (point A)	Angle with north-south line is $90° - 11.82° = 78.18°$ Ship bearing is: $N\ 78.18°\ W$	

Trig Topics – Part 2

Graphs of Trig Functions

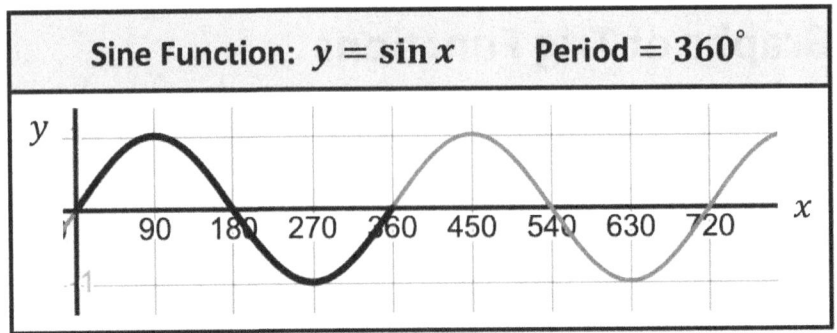

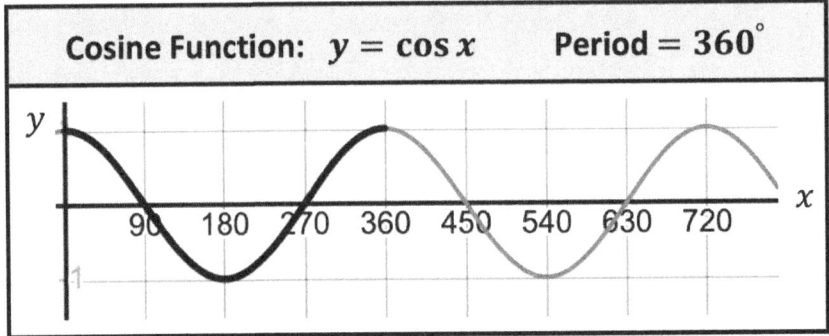

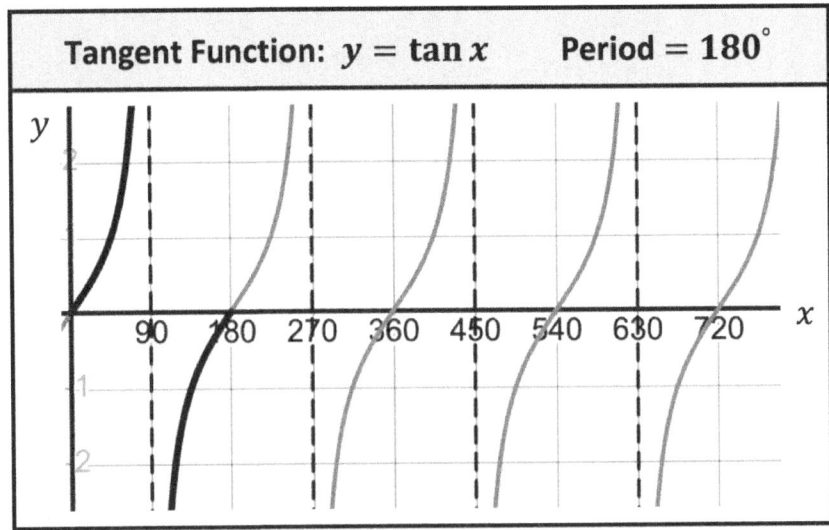

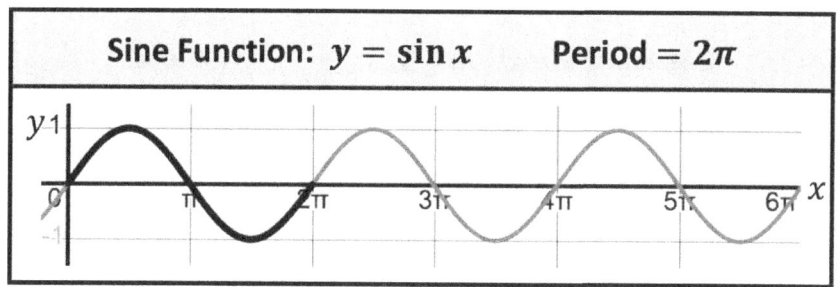

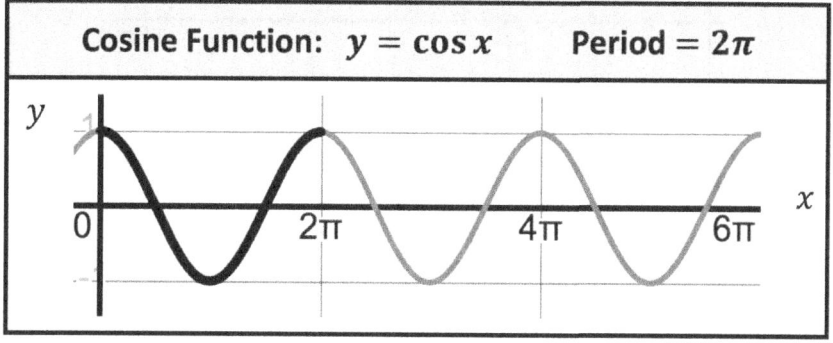

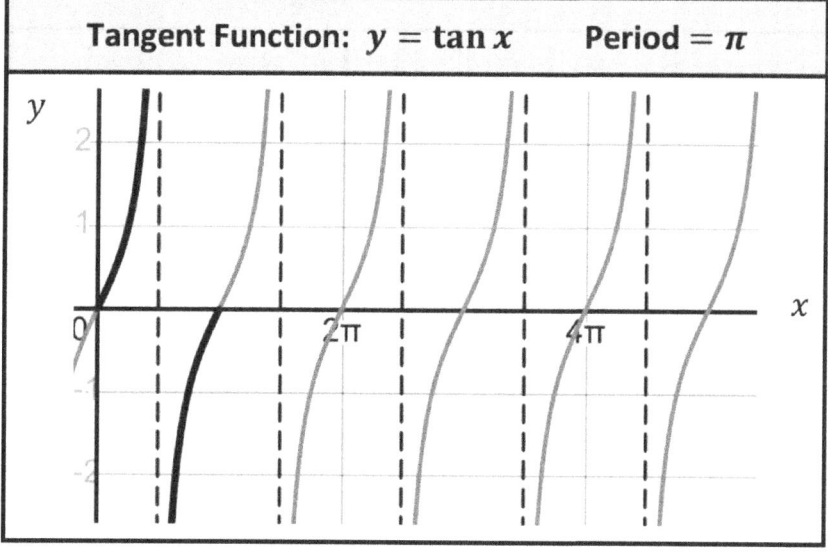

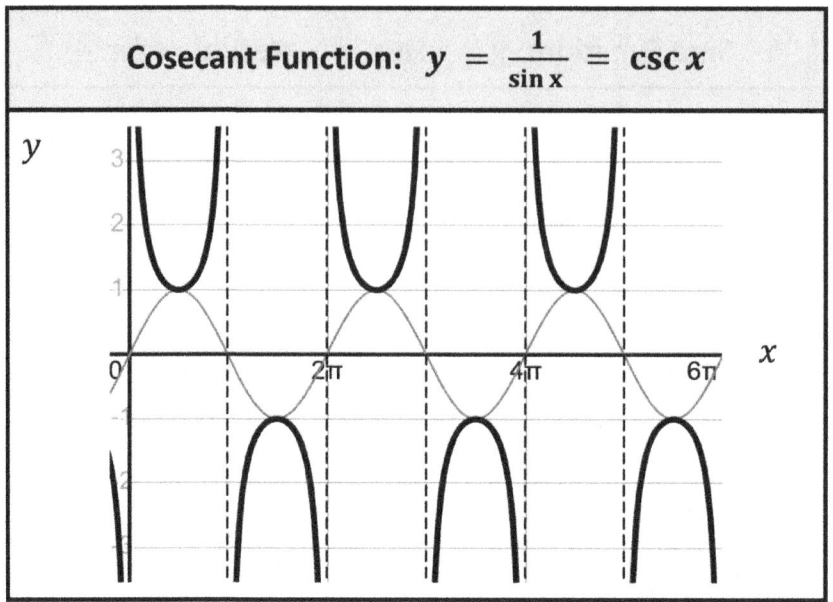

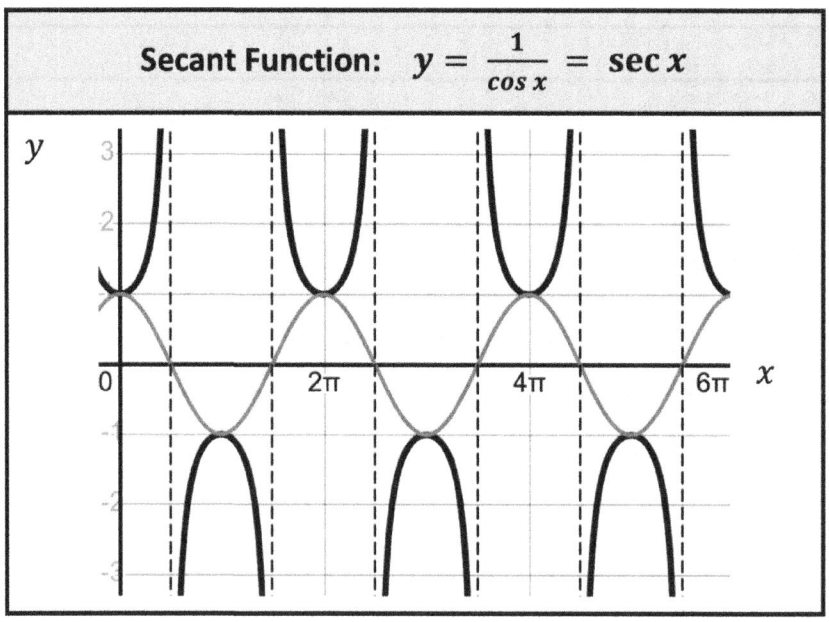

Graphs of Inverse Trig Functions

Inverse Sine Function: $y = \sin^{-1} x$

Domain: $[-1, 1]$ Range: $\left[-\dfrac{\pi}{2}, \dfrac{\pi}{2}\right]$ Q1 & Q4

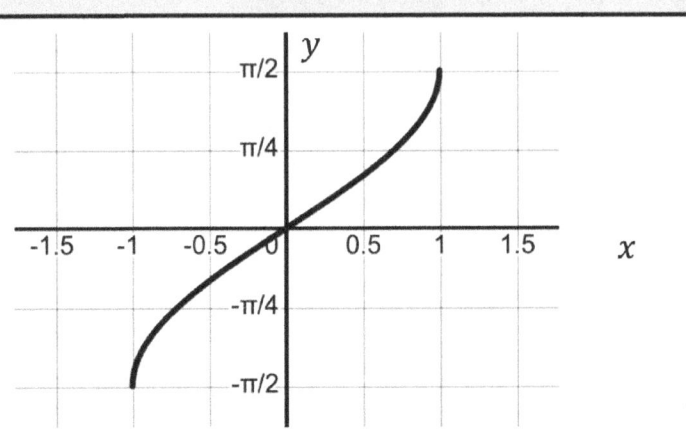

Inverse Cosine Function: $y = \cos^{-1} x$

Domain: $[-1, 1]$ Range: $[0, \pi]$ Q1 & Q2

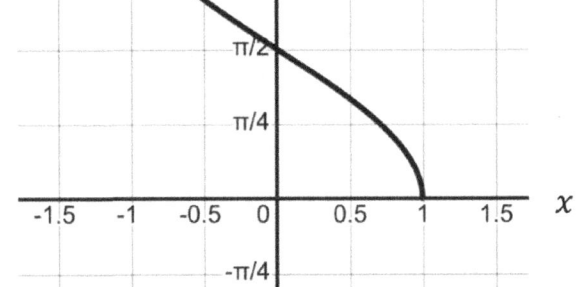

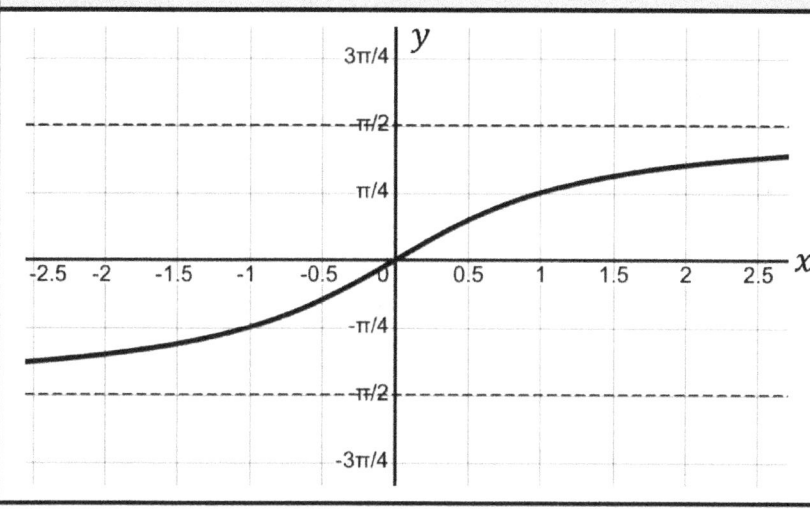

Translations of Trig Functions

Translations of Trig Functions

The general format for a trigonometric function is:

$$y = a \cdot trig(b(x+c)) + d$$

Where:
- a = Vertical stretch or compression (amplitude)
- b = Horizontal stretch or compression
- c = Horizontal shift
- d = Vertical shift

And:
- $trig = \sin, \cos, \tan, \sec, \csc, \cot, \sin^{-1}, \ldots$

Translations of Trig Functions – Ex. 1

Graph the function: $y = 2 \cdot sin(x + 30°)$

Identify the translations	V. Stretch $= 2$ (Amplitude) H. Shift left $= 30°$

Create an extended T-table to help with graphing.

$x - 30$	x	y	$2y$
-30	0	0	0
60	90	1	2
150	180	0	0
240	270	-1	-2
330	360	0	0

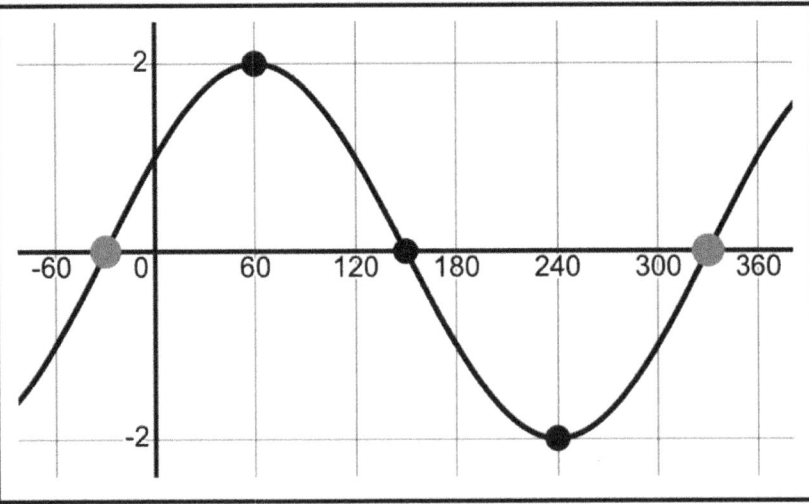

Translations of Trig Functions – Ex. 2
Graph the function: $y = \sin(2x + 30°) + 1$

Rewrite	$y = \sin\left(2(x + 15°)\right) + 1$
Identify the translations	V. Shift up 1 (Centerline: $y = 1$) H. Compression by ½ (Period = π) H. Shift left by = $15°$

Create an extended T-table to help with graphing.

$\left(\frac{1}{2}\right)x - 15$	x	y	$y + 1$
−15	0	0	1
30	90	1	2
75	180	0	1
120	270	−1	0
165	360	0	1

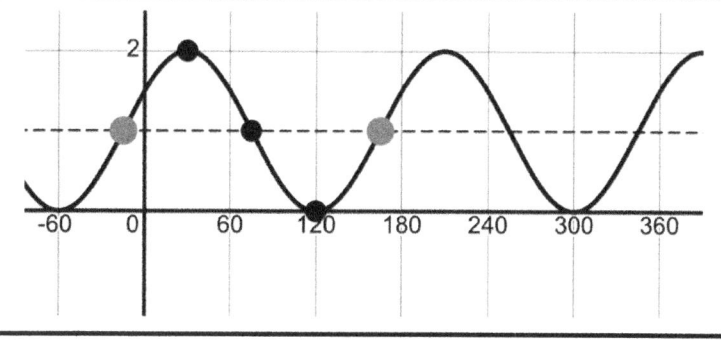

Translations of Trig Functions – Ex. 3

Graph the function:	$y = 3\cos(2x - \pi) - 1$
Rewrite	$y = 3\cos\left(2\left(x - \frac{\pi}{2}\right)\right) - 1$
Identify the translations	V. Stretch by 3 (Amplitude = 3) V. Shift down 1 (Centerline: $y = -1$) H. Compression by ½ (Period = π) H. Shift right by $\frac{\pi}{2}$

$\left(\frac{1}{2}\right)x + \frac{\pi}{2}$	x	y	$3y - 1$
$\frac{\pi}{2}$	0	1	2
$\frac{3\pi}{4}$	$\frac{\pi}{2}$	0	-1
π	π	-1	-4
$\frac{5\pi}{4}$	$\frac{3\pi}{2}$	0	-1
$\frac{3\pi}{2}$	2π	1	2

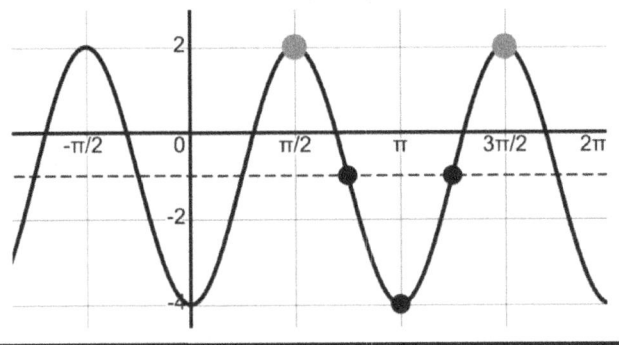

Translations of Trig Functions – Ex. 4a

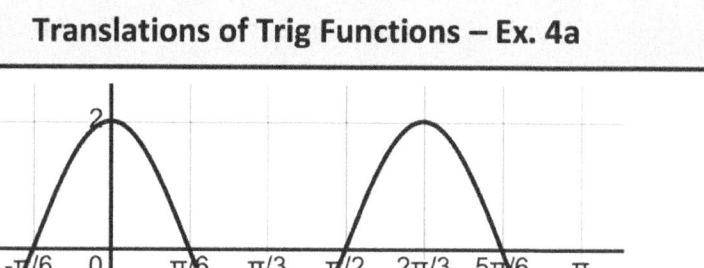

Write the function for the above graph as a sine function and as a cosine function.

Sine Funct. Start: $-\dfrac{\pi}{6}$ End: $\dfrac{\pi}{2}$ Period $= \dfrac{2\pi}{3}$	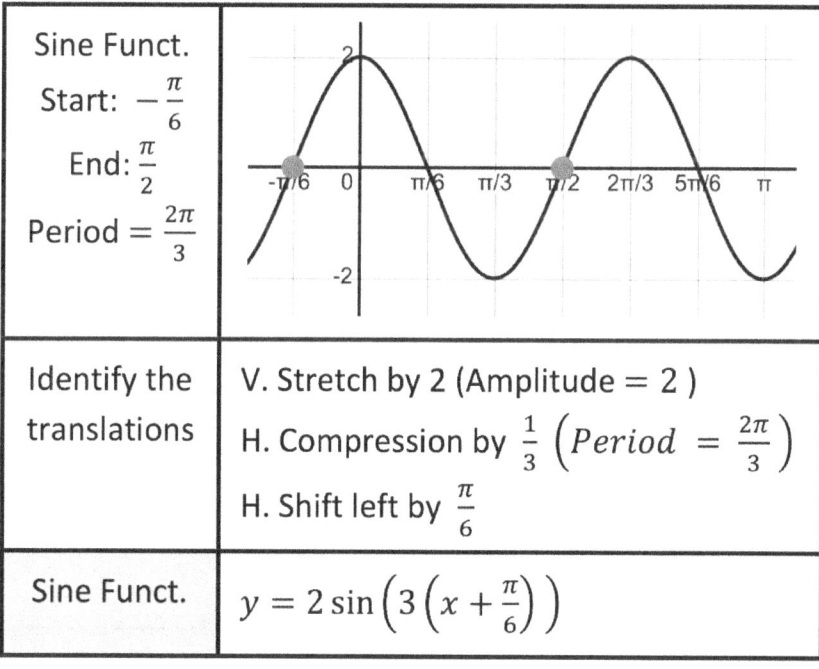
Identify the translations	V. Stretch by 2 (Amplitude $= 2$) H. Compression by $\dfrac{1}{3}$ $\left(Period = \dfrac{2\pi}{3}\right)$ H. Shift left by $\dfrac{\pi}{6}$
Sine Funct.	$y = 2\sin\left(3\left(x + \dfrac{\pi}{6}\right)\right)$

Translations of Trig Functions – Ex. 4b

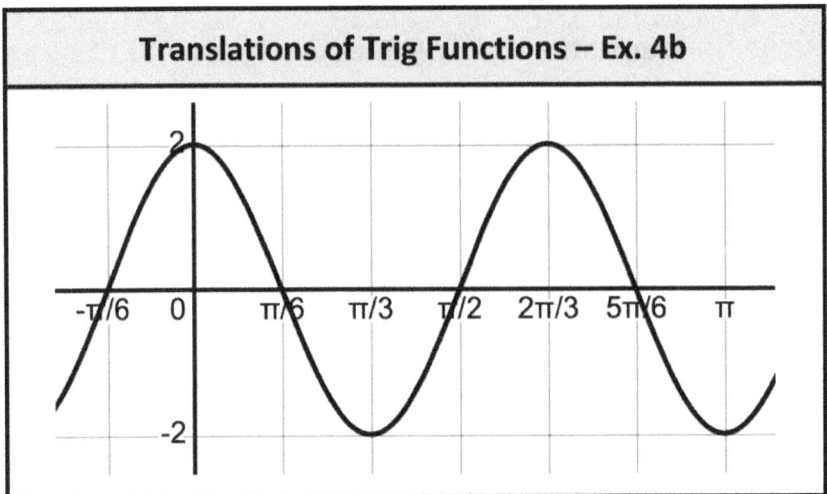

Write the function for the above graph as a sine function and as a cosine function.

Cos. Funct. Start: 0 End: $\frac{2\pi}{3}$ Period $= \frac{2\pi}{3}$	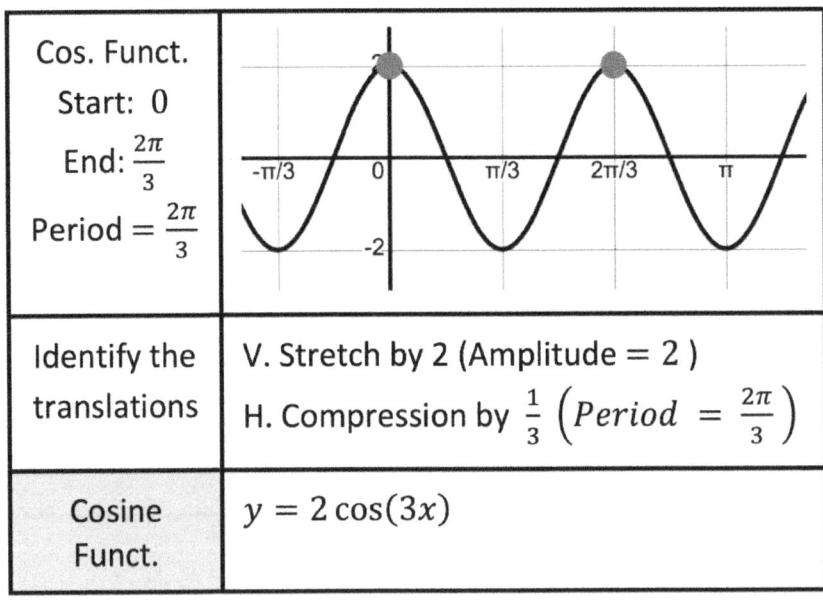	
Identify the translations	V. Stretch by 2 (Amplitude = 2) H. Compression by $\frac{1}{3}$ $\left(Period = \frac{2\pi}{3}\right)$	
Cosine Funct.	$y = 2\cos(3x)$	

Translations of Trig Functions – Ex. 10c

From a platform (at $t = 0$) you get into your seat on a ferris wheel. It takes 60 seconds to reach the top, which is 220 ft. above the ground. It takes 160 sec. to make make one revolution. Diameter of the ferris wheel is 200 ft. Find: Equation for your distance from the ground as a function of time.

Modeled with sin.	$y = a \sin(b(x+c)) + d$ $y = 100 \sin\left(\frac{\pi}{80}(x-20)\right) + 110$

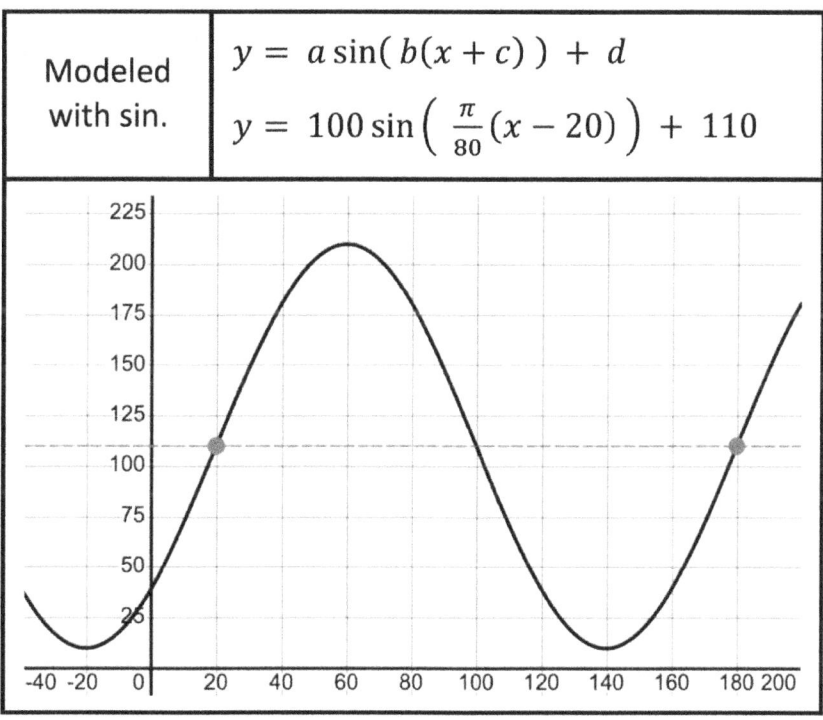

Trig Identities

Trig. Definitions and Identities
This section contains lists of trigonometric definitions and identities. A copy of the unit circle is also included.

LAST PAGE	<ul><li>The last page contains a summary list for quick reference.</li><li>Students may find it useful to print a copy of the last page or take a picture of it, for quick reference.</li></ul>

Definition of Six Trig Functions

Right Triangle Definitions

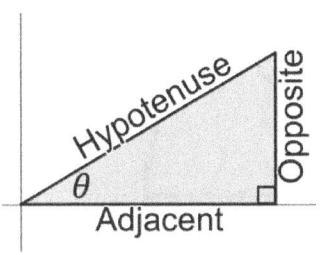

$\sin\theta = \dfrac{opp}{hyp}$ $\csc\theta = \dfrac{1}{\sin\theta}$

$\cos\theta = \dfrac{adj}{hyp}$ $\sec\theta = \dfrac{1}{\cos\theta}$

$\tan\theta = \dfrac{opp}{adj}$ $\cot\theta = \dfrac{1}{\tan\theta}$

Circular Function Definitions

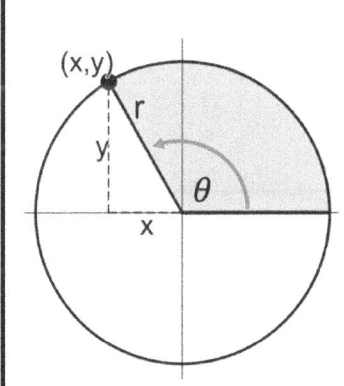

$r = \sqrt{x^2 + y^2}$

$\sin\theta = \dfrac{y}{r}$

$\cos\theta = \dfrac{x}{r}$

$\tan\theta = \dfrac{y}{x}$

The Unit Circle

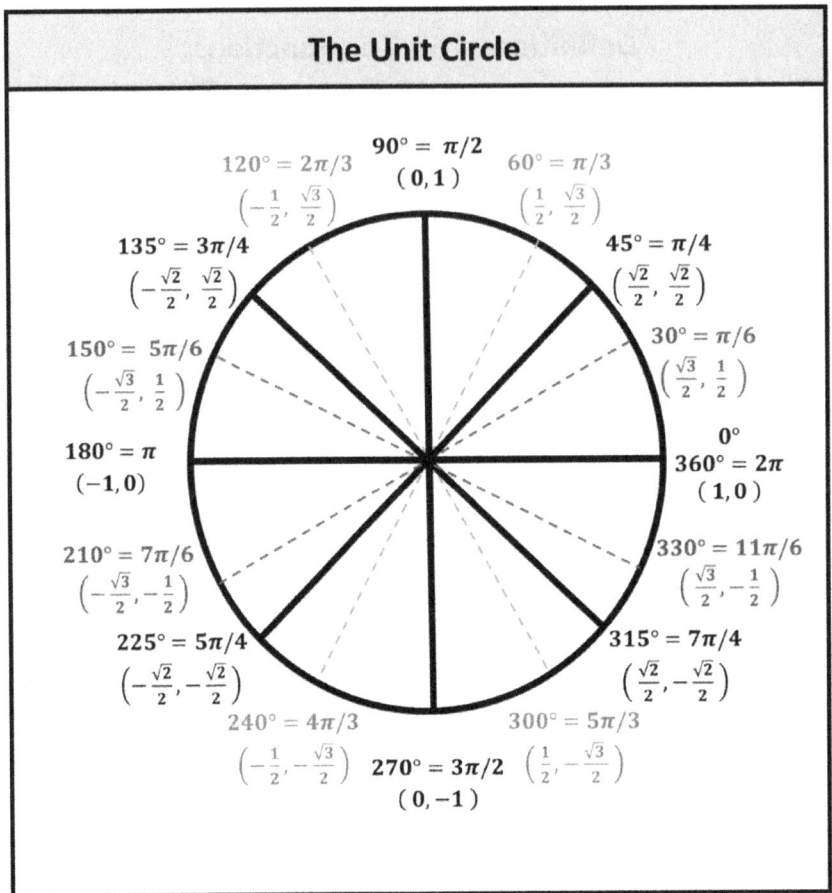

Trig. Function	When $r = 1$
$x = r \cos \theta$	$x = \cos \theta$
$y = r \sin \theta$	$y = r \sin \theta$

Pythagorean Identities
$$\sin^2 x + \cos^2 x = 1$$
Note: The other Pythagorean Identities can be easily derived from this identity. Just memorize this one!!!

Divide both sides By $\sin^2 x$	$\sin^2 x + \cos^2 x = 1$ $1 + \cot^2 x = \csc^2 x$

Divide both sides By $\cos^2 x$	$\sin^2 x + \cos^2 x = 1$ $\tan^2 x + 1 = \sec^2 x$

Confunction Identities

$\sin\left(\frac{\pi}{2} - x\right) = \cos x$	$\cos\left(\frac{\pi}{2} - x\right) = \sin x$
$\csc\left(\frac{\pi}{2} - x\right) = \sec x$	$\tan\left(\frac{\pi}{2} - x\right) = \cot x$
$\sec\left(\frac{\pi}{2} - x\right) = \csc x$	$\cot\left(\frac{\pi}{2} - x\right) = \tan x$

Even/Odd Identities

$\sin(-x) = -\sin x$	$\cos(-x) = \cos x$
$\csc(-x) = -\csc x$	$\tan(-x) = -\tan x$
$\sec(-x) = \sec x$	$\cot(-x) = -\cot x$

It will be helpful to
visualize the above identities
on a unit circle.

Trig. Definitions

$\cos\theta = \dfrac{x}{r}$	$\sin\theta = \dfrac{y}{r}$	$\tan\theta = \dfrac{y}{x}$
$\sec\theta = \dfrac{1}{\cos\theta}$	$\csc\theta = \dfrac{1}{\sin\theta}$	$\cot\theta = \dfrac{1}{\tan\theta}$

Pythagorean Identity

$$\sin^2 x + \cos^2 x = 1 \quad \text{(derive others)}$$

Confunction Identities

$\sin\left(\dfrac{\pi}{2} - x\right) = \cos x$	$\cos\left(\dfrac{\pi}{2} - x\right) = \sin x$
$\csc\left(\dfrac{\pi}{2} - x\right) = \sec x$	$\tan\left(\dfrac{\pi}{2} - x\right) = \cot x$
$\sec\left(\dfrac{\pi}{2} - x\right) = \csc x$	$\cot\left(\dfrac{\pi}{2} - x\right) = \tan x$

Even/Odd Identities

$\sin(-x) = -\sin x$	$\cos(-x) = \cos x$
$\csc(-x) = -\csc x$	$\tan(-x) = -\tan x$
$\sec(-x) = \sec x$	$\cot(-x) = -\cot x$

Trig Formulas

Trig. Formulas
This section contains lists of trigonometric formulas.

LAST PAGE	• The last page contains a summary list for quick reference. • Students may find it useful to print a copy of the last page or take a picture of it, for quick reference.

Sum and Difference Formulas

$\sin(u \pm v)$	$= \sin u \cos v \pm \cos u \sin v$
$\cos(u \pm v)$	$= \cos u \cos v \mp \sin u \sin v$
$\tan(u \pm v)$	$= \dfrac{\tan u \pm \tan v}{1 \mp \tan u \tan v}$

Double-Angle Formulas

$\sin 2u$	$= 2 \sin u \cos u$
$\cos 2u$	$= \cos^2 u - \sin^2 u$
	$= 2 \cos^2 u - 1 = 1 - 2 \sin^2 u$
$\tan 2u$	$= \dfrac{2 \tan u}{1 - \tan^2 u}$

Power-Reducing Formulas

$\sin^2 u = \dfrac{1 - \cos 2u}{2}$	$\cos^2 u = \dfrac{1 + \cos 2u}{2}$
$\tan^2 u = \dfrac{1 - \cos 2u}{1 + \cos 2u}$	

Sum-to-Product Formulas

$$\sin u + \sin v = 2 \sin\left(\frac{u+v}{2}\right) \cdot \cos\left(\frac{u-v}{2}\right)$$

$$\sin u - \sin v = 2 \cos\left(\frac{u+v}{2}\right) \cdot \sin\left(\frac{u-v}{2}\right)$$

$$\cos u + \cos v = 2 \cos\left(\frac{u+v}{2}\right) \cdot \cos\left(\frac{u-v}{2}\right)$$

$$\cos u - \cos v = -2 \sin\left(\frac{u+v}{2}\right) \cdot \sin\left(\frac{u-v}{2}\right)$$

Product-to-Sum Formulas

$$\sin u \sin v = \frac{1}{2}\left[\cos(u-v) - \cos(u+v)\right]$$

$$\cos u \cos v = \frac{1}{2}\left[\cos(u-v) + \cos(u+v)\right]$$

$$\sin u \cos v = \frac{1}{2}\left[\sin(u+v) + \sin(u-v)\right]$$

$$\cos u \sin v = \frac{1}{2}\left[\sin(u+v) - \sin(u-v)\right]$$

Sum and Difference Formulas
$\sin(u \pm v) = \sin u \cos v \pm \cos u \sin v$
$\cos(u \pm v) = \cos u \cos v \mp \sin u \sin v$
$\tan(u \pm v) = \dfrac{\tan u \pm \tan v}{1 \mp \tan u \tan v}$

Double Angle	Power-Reducing
$\sin 2u = 2 \sin u \cos u$	$\sin^2 u = \dfrac{1 - \cos 2u}{2}$
$\cos 2u = \cos^2 u - \sin^2 u$ $= 2\cos^2 u - 1 = 1 - 2\sin^2 u$	$\cos^2 u = \dfrac{1 + \cos 2u}{2}$
$\tan 2u = \dfrac{2 \tan u}{1 - \tan^2 u}$	$\tan^2 u = \dfrac{1 - \cos 2u}{1 + \cos 2u}$

Sum-to-Product Formulas
$\sin u + \sin v = 2 \sin\left(\dfrac{u+v}{2}\right) \cdot \cos\left(\dfrac{u-v}{2}\right)$
$\sin u - \sin v = 2 \cos\left(\dfrac{u+v}{2}\right) \cdot \sin\left(\dfrac{u-v}{2}\right)$
$\cos u + \cos v = 2 \cos\left(\dfrac{u+v}{2}\right) \cdot \cos\left(\dfrac{u-v}{2}\right)$
$\cos u - \cos v = -2 \sin\left(\dfrac{u+v}{2}\right) \cdot \sin\left(\dfrac{u-v}{2}\right)$

Product-to-Sum Formulas
$\sin u \sin v = \dfrac{1}{2}[\cos(u-v) - \cos(u+v)]$
$\cos u \cos v = \dfrac{1}{2}[\cos(u-v) + \cos(u+v)]$
$\sin u \cos v = \dfrac{1}{2}[\sin(u+v) + \sin(u-v)]$
$\cos u \sin v = \dfrac{1}{2}[\sin(u+v) - \sin(u-v)]$

Using Trig Identities

Using Trig. Identities To Evaluate a Function – Ex. 1

Given: $\sec u = -\dfrac{3}{2}$ and $\tan u > 0$

Find: All six trigonometric functions.

Identify Quadrant	Negative secant (or cosine) → Q2, Q3
	Positive tangent → Q1, Q3
	Therefore: Angle u is in Q3

Make a Sketch in Q1	$\sec u = \dfrac{1}{\cos u}$ $\cos u = \dfrac{2}{3} = \dfrac{x}{r}$	(triangle with hypotenuse 3, adjacent 2, opposite y)

Find y	$x^2 + y^2 = r^2$
	$y = \sqrt{r^2 - x^2} = \sqrt{3^2 - 2^2} = \sqrt{5}$

Six Trig. Functions. Note: For Q3 $(x, y) = (-, -)$

$\cos u = -\dfrac{2}{3}$	$\sin u = -\dfrac{\sqrt{5}}{3}$	$\tan u = \dfrac{\sqrt{5}}{2}$
$\sec u = -\dfrac{3}{2}$	$\csc u = -\dfrac{\sqrt{5}}{3}$	$\cot u = \dfrac{2}{\sqrt{5}}$

Using Trig. Identities To Simplify an Expression – Ex. 2	
Simplify: $2\sin x \cos^2 x - 2\sin x$	

Factor	$2\sin x \cos^2 x - 2\sin x$ $2\sin x (\cos^2 x - 1)$
Substitute	$2\sin x (\cos^2 x - (\sin^2 x + \cos^2 x))$
Simplify	$2\sin x (\cos^2 x - \sin^2 x - \cos^2 x)$ $2\sin x (-\sin^2 x)$ $-2\sin^3 x$

Verify a Trig. Identity – Ex. 3
Verify: $\dfrac{\sin u}{1+\cos u} + \dfrac{\cos u}{\sin u} = \csc u$

$\dfrac{\sin u}{1+\cos u} + \dfrac{\cos u}{\sin u}$	$= \csc u$
$\dfrac{\sin^2 u + \cos u(1+\cos u)}{(1+\cos u)\sin u}$	
$\dfrac{\sin^2 u + \cos u + \cos^2 u}{(1+\cos u)\sin u}$	
$\dfrac{1+\cos u}{(1+\cos u)\sin u}$	
$\dfrac{1}{\sin u}$	
$\csc u$	$= \csc u$

Factor a Trig. Expression – Ex. 4
Factor: $\csc^2 u - 4\cot u - 6$

Consider Pythagorean Identities	$\sin^2 u + \cos^2 u = 1$ $1 + \cot^2 u = \csc^2 u$
Substitute	$\csc^2 u - 4\cot u - 6$ Original Eqn. $(1 + \cot^2 u) - 4\cot u - 6$
Simplify	$\cot^2 u - 4\cot u - 5$
Factor	$(\cot u + 1)(\cot u - 5)$

Simplify a Trig. Expression – Ex. 5

Simplify: $\cos x + \tan x \sin x$

You can always convert everything to sin and cos and see what happens!	$\cos x + \tan x \sin x$ $\cos x + \left(\dfrac{\sin x}{\cos x}\right) \sin x$ $\cos x + \left(\dfrac{\sin^2 x}{\cos x}\right)$
Write as a single fraction	$\dfrac{\cos^2 x + \sin^2 x}{\cos x}$
Simplify	$\dfrac{1}{\cos x}$
Rewrite	$\sec x$

Rewrite a Trig. Expression – Ex. 6

Given: $\dfrac{1}{1 - \sin x}$

Rewrite so the expression does not include fractions.

Reformat fraction so it has one term in denominator.	$\dfrac{1}{1 - \sin x}\left(\dfrac{1 + \sin x}{1 + \sin x}\right)$ $\dfrac{1 + \sin x}{1 - \sin^2 x}$ $\dfrac{1 + \sin x}{1 - (1 - \cos^2 x)}$ $\dfrac{1 + \sin x}{\cos^2 x}$
Simplify	$\dfrac{1 + \sin x}{\cos^2 x}$ $\dfrac{1}{\cos^2 x} + \dfrac{\sin x}{\cos^2 x}$ $\dfrac{1}{\cos^2 x} + \left(\dfrac{1}{\cos x}\right)\left(\dfrac{\sin x}{\cos x}\right)$ $\sec^2 x + \sec x \tan x$
Factor (Just for fun!)	$\sec x (\sec x + \tan x)$

Verify a Trig. Identity – Ex. 7
Verify: $\dfrac{\sec^2 x - 1}{\sec^2 x} = \sin^2 x$

$\dfrac{\sec^2 x - 1}{\sec^2 x}$	$= \sin^2 x$
$\dfrac{\sec^2 x}{\sec^2 x} - \dfrac{1}{\sec^2 x}$	
$1 - \cos^2 x$	
$\sin^2 x$	$= \sin^2 x$

Verify a Trig. Identity – Ex. 8
Verify: $\dfrac{1}{1-\cos x} + \dfrac{1}{1+\cos x} = 2\csc^2 x$

$\dfrac{1}{1-\cos x} + \dfrac{1}{1+\cos x}$	$= 2\csc^2 x$
$\dfrac{(1+\cos x) + (1-\cos x)}{(1-\cos x)(1+\cos x)}$	
$\dfrac{2}{1-\cos^2 x}$	
$\dfrac{2}{\sin^2 x}$	
$2\left(\dfrac{1}{\sin^2 x}\right)$	
$2\csc^2 x$	$= 2\csc^2 x$

Verify a Trig. Identity – Ex. 9
Verify: $(\tan^2 x + 1)(\sin^2 x - 1) = -1$

$(\tan^2 x + 1)(\sin^2 x - 1)$	$= -1$
$\left(\dfrac{1}{\cos^2 x}\right)(-\cos^2 x)$	
$\dfrac{-\cos^2 x}{\cos^2 x}$	
-1	$= -1$

Recall: Pythagorean Identities	$\sin^2 x + \cos^2 x = 1$
	$\dfrac{\sin^2 x}{\cos^2 x} + \dfrac{\cos^2 x}{\cos^2 x} = \dfrac{1}{\cos^2 x}$
	$\tan^2 x + 1 = \sec^2 x$
	$\tan^2 x + 1 = \dfrac{1}{\cos^2 x}$

Verify a Trig. Identity – Ex. 10
Verify: $\tan x + \cot x = \sec x \csc x$

$\tan x + \cot x$	$= \sec x \csc x$
$\dfrac{\sin x}{\cos x} + \dfrac{\cos x}{\sin x}$	
$\dfrac{\sin^2 x}{\cos x \sin x} + \dfrac{\cos^2 x}{\cos x \sin x}$	
$\dfrac{\sin^2 x + \cos^2 x}{\cos x \sin x}$	
$\dfrac{1}{\cos x \sin x}$	
$\left(\dfrac{1}{\cos x}\right)\left(\dfrac{1}{\sin x}\right)$	
$\sec x \csc x$	$= \sec x \csc x$

Verify a Trig. Identity – Ex. 11

Verify: $\sec x - \tan x = \dfrac{\cos x}{1 + \sin x}$

$\sec x - \tan x$	$=$	$\dfrac{\cos x}{1 + \sin x}$
		$\dfrac{\cos x}{1 + \sin x} \left(\dfrac{1 - \sin x}{1 - \sin x} \right)$
		$\dfrac{\cos x - \cos x \sin x}{1 - \sin^2 x}$
		$\dfrac{\cos x - \cos x \sin x}{\cos^2 x}$
		$\dfrac{\cos x}{\cos^2 x} - \dfrac{\cos x \sin x}{\cos^2 x}$
		$\dfrac{1}{\cos x} - \dfrac{\sin x}{\cos x}$
$\sec x - \tan x$	$=$	$\sec x - \tan x$

Solving Trig Equations

	Solving Trig. Equations – Ex. 1
	Given: $2 \sin x - 4 = 5$ Solve for x

Use algebra	$2 \sin x - 4 = 5$ $2 \sin x = 1$ $\sin x = \dfrac{1}{2}$ $x = \sin^{-1}\left(\dfrac{1}{2}\right)$
Use inverse trig function	$x = \sin^{-1}\left(\dfrac{1}{2}\right)$ " x is the angle whose sine is $\dfrac{1}{2}$ "
Visualize on a unit circle.	There are two angles that have a sine of $\dfrac{1}{2}$
Two solutions	$x = \dfrac{\pi}{6}, \dfrac{5\pi}{6}$
Infinite solutions	$x = \dfrac{\pi}{6} + n(2\pi)$ and $x = \dfrac{5\pi}{6} + n(2\pi)$; $n = 0, 1, 2, ...$

	Solving Trig. Equations – Ex. 2
	Given: $3\sin x + \sqrt{2} = \sin x$ Solve for x

Use algebra	$3\sin x + \sqrt{2} = \sin x$ $2\sin x = -\sqrt{2}$ $\sin x = -\dfrac{\sqrt{2}}{2}$
Use inverse trig function	$x = \sin^{-1}\left(-\dfrac{\sqrt{2}}{2}\right)$ " x is the angle whose sine is $-\dfrac{\sqrt{2}}{2}$ "
Visualize on a unit circle.	There are two angles that have that sine
Two solutions	$x = \dfrac{5\pi}{4}, \dfrac{7\pi}{4}$
Infinite solutions	$x = \dfrac{5\pi}{4} + n(2\pi)$ and $x = \dfrac{7\pi}{4} + n(2\pi)$; $n = 0, 1, 2, \ldots$

Solving Trig. Equations – Ex. 3	
Given: $6\tan^2 x - 2 = 0$;	Solve for x

Use algebra	$6\tan^2 x - 2 = 0$ $\tan^2 x = \dfrac{2}{6} = \dfrac{1}{3}$ $\tan x = \pm\sqrt{\dfrac{1}{3}} = \pm\dfrac{1}{\sqrt{3}} = \pm\dfrac{\sqrt{3}}{3}$
Use inverse trig function	$x = \tan^{-1}\left(\dfrac{\sqrt{3}}{3}\right)$ or $x = \tan^{-1}\left(-\dfrac{\sqrt{3}}{3}\right)$
Visualize on a unit circle.	On the unit circle, there are four angles that have those tangents.
Four solutions	$x = \dfrac{\pi}{3}, \dfrac{2\pi}{3}, \dfrac{4\pi}{3}, \dfrac{5\pi}{3}$
Infinite solutions	$x = \dfrac{\pi}{3} + n(\pi)$ and $x = \dfrac{2\pi}{4} + n(\pi)$; $n = 0, 1, 2, \ldots$

Solving Trig. Equations – Ex. 4

Given: $\cot x \cos^2 x = 3 \cot x$; Solve for x

Use Algebra. Factor.	$\cot x \cos^2 x = 3 \cot x$ $\cot x \cos^2 x - 3 \cot x = 0$ $\cot x (\cos^2 x - 3) = 0$	
	Note: Cannot divide by $\cot x$ because it may equal zero.	
Several solutions	$\cot x = 0$ $\dfrac{\cos x}{\sin x} = 0$ $\cos x = 0$ $x = \dfrac{\pi}{2}, \dfrac{3\pi}{2}$	$\cos^2 x - 3 = 0$ $\cos x = \pm\sqrt{3}$ But: $-1 \leq \cos x \leq 1$ Extraneous sol'n
Visualize on a unit circle.		
Several solutions	$x = \dfrac{\pi}{2}, \dfrac{3\pi}{2}$	
Infinite solutions	$x = \dfrac{\pi}{2} + n(\pi)$	

Solving Trig. Equations – Ex. 5
Find all solutions for: $2\sin^2 x - \sin x = 1$
In the interval: $[0, 2\pi)$

Set = 0 Then, factor.	$2\sin^2 x - \sin x = 1$ $2\sin^2 x - \sin x - 1 = 0$ $(2\sin x + 1)(\sin x - 1) = 0$	
Several solutions	$2\sin x + 1 = 0$ $\sin x = -\dfrac{1}{2}$ $x = \dfrac{7\pi}{6}, \dfrac{11\pi}{6}$	$\sin x - 1 = 0$ $\sin x = 1$ $x = \dfrac{\pi}{2}$
Visualize on a unit circle.		
Several solutions	$x = \dfrac{\pi}{2}, \dfrac{7\pi}{6}, \dfrac{11\pi}{6}$	
Solutions in $[0, 2\pi)$	$x = \dfrac{\pi}{2}, \dfrac{7\pi}{6}, \dfrac{11\pi}{6}$	

Solving Trig. Equations – Ex. 6

Find all solutions for: $\cos x - \sin x + 1 = 0$
In the interval: $[0, 2\pi)$

Rearrange, then square both sides.	$\cos x - \sin x + 1 = 0$ $\cos x + 1 = \sin x$ $\cos^2 x + 2\cos x + 1 = \sin^2 x$ Note: This may cause extraneous soln's
Simplify and Factor	$\cos^2 x + 2\cos x + 1 = 1 - \cos^2 x$ $2\cos^2 x + 2\cos x = 0$ $2\cos x (\cos x + 1) = 0$
Several solutions	$2\cos x = 0$ $\qquad$ $\cos x + 1 = 0$ $\cos x = 0$ $\qquad$ $\cos x = -1$ $x = \frac{\pi}{2}, \frac{3\pi}{2}$ $\qquad$ $x = \pi$
Visualize on a unit circle.	
Check for Extraneous Sol'ns. Try them in original eqn.	
Solutions in $[0, 2\pi)$	$x = \frac{\pi}{2}, \pi$ $\qquad$ $\frac{3\pi}{2}$ is an extraneous sol'n

	Solving Trig. Equations – Ex. 7
	Find all solutions for: $4 \cos 3x - 2 = 0$

Use Algebra	$4 \cos 3x - 2 = 0$ $\cos(3x) = \frac{2}{4} = \frac{1}{2}$	
Visualize Unit Circle	$\cos(3x) = \frac{1}{2}$ $(3x) = \frac{\pi}{3}, \frac{5\pi}{3}$	
Solve for x	$3x = \frac{\pi}{3} + n(2\pi)$ $x = \frac{\pi}{9} + \frac{2n\pi}{3}$	$3x = \frac{5\pi}{3} + n(2\pi)$ $x = \frac{5\pi}{9} + \frac{2n\pi}{3}$
All Solutions	$x = \frac{\pi}{9} + \frac{2n\pi}{3}$ or	$x = \frac{5\pi}{9} + \frac{2n\pi}{3}$

	Solving Trig. Equations – Ex. 8
	Find all solutions for: $7 \tan \frac{x}{2} + 7 = 0$

Use Algebra	$7 \tan \frac{x}{2} + 7 = 0$ $\tan \frac{x}{2} = -\frac{7}{7} = -1$	
Visualize Unit Circle	$\tan \left(\frac{x}{2}\right) = -1$ $\left(\frac{x}{2}\right) = \frac{3\pi}{4}, \frac{7\pi}{4}$	
Solve for x	$\left(\frac{x}{2}\right) = \frac{3\pi}{4} + n(\pi)$ $x = \frac{3\pi}{2} + 2n\pi$	Note: The period for tangent is π.
All Solutions	$x = \frac{3\pi}{2} + 2n\pi$	

	Solving Trig. Equations – Ex. 9
	Find all solutions for: $2\sec^2 x - 4\tan x = 8$

Simplify, substitute and factor	$2\sec^2 x - 4\tan x = 8$ $\sec^2 x - 2\tan x = 4$ $(1 + \tan^2 x) - 2\tan x - 4 = 0$ $\tan^2 x - 2\tan x - 3 = 0$ $(\tan x - 3)(\tan x + 1) = 0$
Several solutions	$\tan x - 3 = 0$ $\tan x = 3$ $x = \tan^{-1}(3)$	$\tan x + 1 = 0$ $\tan x = -1$ $x = \dfrac{3\pi}{4}, \dfrac{7\pi}{4}$
Visualize Unit Circle		
All Solutions	$x = \tan^{-1}(3) + n\pi$ $\quad$ and $\quad$ $x = \dfrac{3\pi}{4} + n\pi$

Using Trig Formulas

Using Trig. Formulas – Ex. 1
Find the exact value of $\sin 75°$

Use Sum & Difference Formulas	$\sin(75°)$ $\sin(30° + 45°)$ $\sin 30° \cos 45° +$ $\cos 30° \sin 45°$ $\left(\frac{1}{2}\right)\left(\frac{\sqrt{2}}{2}\right) + \left(\frac{\sqrt{3}}{2}\right)\left(\frac{\sqrt{2}}{2}\right)$ $\frac{\sqrt{2}}{4} + \frac{\sqrt{6}}{4}$ $\frac{\sqrt{2} + \sqrt{6}}{4}$

	Using Trig. Formulas – Ex. 2
Find the exact value of $\cos\dfrac{\pi}{12}$	

Represent the angle with 2 angles	$\dfrac{\pi}{12} = \dfrac{4\pi}{12} - \dfrac{3\pi}{12}$ $\dfrac{\pi}{12} = \dfrac{\pi}{3} - \dfrac{\pi}{4}$
Use Sum & Difference Formulas	$\cos\dfrac{\pi}{12}$ $\cos\left(\dfrac{\pi}{3} - \dfrac{\pi}{4}\right)$ $\cos\dfrac{\pi}{3}\cos\dfrac{\pi}{4} + \sin\dfrac{\pi}{3}\sin\dfrac{\pi}{4}$ $\left(\dfrac{1}{2}\right)\left(\dfrac{\sqrt{2}}{2}\right) + \left(\dfrac{\sqrt{3}}{2}\right)\left(\dfrac{\sqrt{2}}{2}\right)$ $\dfrac{\sqrt{2}}{4} + \dfrac{\sqrt{6}}{4}$ $\dfrac{\sqrt{2} + \sqrt{6}}{4}$

Using Trig. Formulas – Ex. 3
Find the exact value of: $$\cos 100° \cos 10° + \sin 100° \sin 10°$$

Note ...	$90° = 100° - 10°$
Note the pattern	$\cos 100° \cos 10° + \sin 100° \sin 10°$ $\cos(100° - 10°)$ $\cos 90°$
Answer	$\cos 90° = 1$

Using Trig. Formulas – Ex. 4
Evaluate: $\cos\left(\arctan 1 + \arccos\dfrac{1}{2}\right)$

Find the angles. Use default domains.	$A = \arctan 1 = 45°$ $B = \arccos\dfrac{1}{2} = 60°$
Use Sum & Difference Formulas	$\cos\left(\arctan 1 + \arccos\dfrac{1}{2}\right)$ $\cos(45° + 60°)$ $\cos 45° \cos 60° - \sin 45° \sin 60°$ $\left(\dfrac{\sqrt{2}}{2}\right)\left(\dfrac{1}{2}\right) - \left(\dfrac{\sqrt{2}}{2}\right)\left(\dfrac{\sqrt{3}}{2}\right)$ $= \dfrac{\sqrt{2}}{4} - \dfrac{\sqrt{6}}{4}$ $= \dfrac{\sqrt{2} - \sqrt{6}}{4}$

Using Trig. Formulas – Ex. 5
Evaluate: $\cos(\arctan 1 + \arccos x)$

Find the angles. Use default domains.	$A = \arctan 1 = 45°$ $B = \arccos x = ???$
Make a diagram to represent angle B	(triangle with sides 1, $\sqrt{1-x^2}$, x, angle B) $\cos B = x$ $\sin B = \sqrt{1-x^2}$
Use Sum & Difference Formulas	$\cos(\arctan 1 + \arccos x)$ $\cos(A + B)$ $\cos A \cos B - \sin A \sin B$ $\left(\frac{\sqrt{2}}{2}\right)(x) - \left(\frac{\sqrt{2}}{2}\right)\left(\sqrt{1-x^2}\right)$ $= \dfrac{x\sqrt{2}}{2} - \dfrac{\sqrt{2}\sqrt{1-x^2}}{2}$ $= \dfrac{\sqrt{2}}{2}\left(x - \sqrt{1-x^2}\right)$

Using Trig. Formulas – Ex. 6
Prove the identity: $\sin\left(\frac{\pi}{2} - x\right) = \cos x$ Hint: Use Sum and Differences Formulas.

$\sin\left(\frac{\pi}{2} - x\right)$	$= \cos x$
$\sin\frac{\pi}{2}\cos x - \cos\frac{\pi}{2}\sin x$	
$(1)\cos x - (0)\sin x$	
$\cos x$	$= \cos x$

Using Trig. Formulas – Ex. 7

Find all solutions of:

$$\cos\left(x + \frac{\pi}{4}\right) + \cos\left(x - \frac{\pi}{4}\right) = 1$$

On the interval: $[0, 2\pi)$

$\cos\left(x + \frac{\pi}{4}\right) =$	$\cos x \cos\frac{\pi}{4} - \sin x \sin\frac{\pi}{4}$ $\cos x \left(\frac{\sqrt{2}}{2}\right) - \sin x \left(\frac{\sqrt{2}}{2}\right)$
$\cos\left(x - \frac{\pi}{4}\right) =$	$\cos x \cos\frac{\pi}{4} + \sin x \sin\frac{\pi}{4}$ $\cos x \left(\frac{\sqrt{2}}{2}\right) + \sin x \left(\frac{\sqrt{2}}{2}\right)$

$\cos\left(x + \frac{\pi}{4}\right) + \cos\left(x - \frac{\pi}{4}\right) = 1$

$2 \cos x \left(\frac{\sqrt{2}}{2}\right) = 1$

$\cos x = \frac{1}{2}\left(\frac{2}{\sqrt{2}}\right) = \frac{1}{\sqrt{2}} = \frac{\sqrt{2}}{2}$

$x = \cos^{-1}\left(\frac{\sqrt{2}}{2}\right)$ = The angle whose cosine is $\frac{\sqrt{2}}{2}$

$x = \frac{\pi}{4}, \frac{7\pi}{4}$	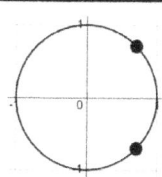

Using Trig. Double Angle Formulas – Ex. 8a

Find all solutions of: $\cos x + \sin 2x = 0$

Rewrite equation in terms of x	$\cos x + \sin 2x = 0$ $\cos x + 2 \sin x \cos x = 0$
Factor	$\cos x (1 + 2 \sin x) = 0$
Solutions	$\cos x = 0$ $1 + 2 \sin x = 0$ $x = \cos^{-1} 0$ $\sin x = -\dfrac{1}{2}$ $x = \dfrac{\pi}{2}, \dfrac{3\pi}{2}$ $x = \dfrac{7\pi}{6}, \dfrac{11\pi}{6}$
All Solutions	$x = \dfrac{\pi}{2} + n\pi$ or $x = \dfrac{7\pi}{6} + 2n\pi$ or $x = \dfrac{11\pi}{6} + 2n\pi$

Using Trig. Double Angle Formulas – Ex. 8b

Find all solutions of: $\cos x + \sin 2x = 0$

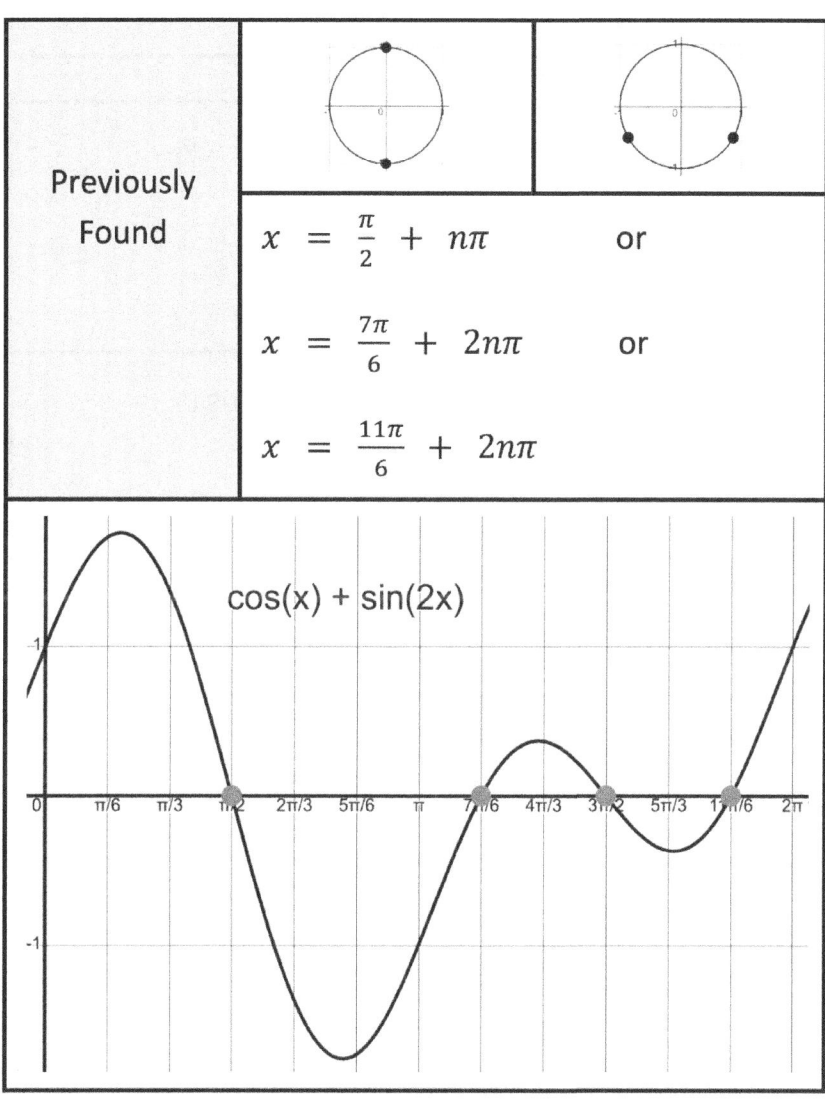

Previously Found

$x = \dfrac{\pi}{2} + n\pi$ or

$x = \dfrac{7\pi}{6} + 2n\pi$ or

$x = \dfrac{11\pi}{6} + 2n\pi$

Using Trig. Double Angle Formulas – Ex. 9

Given: $\cos\theta = \dfrac{3}{5}$; $\dfrac{3\pi}{2} < \theta < 2\pi$

Find: $\sin 2\theta$, $\cos 2\theta$, $\tan 2\theta$

Find the quadrant For θ	$\sin\theta = \dfrac{3}{5}$ $\rightarrow$ Q1 or Q4 $\dfrac{3\pi}{2} < \theta < 2\pi$ $\rightarrow$ Q4 θ is in Q4	
Sketch	(sketch showing angle in Q4, point (3, −4), radius 5)	$\cos\theta = \dfrac{3}{5}$ $\sin\theta = -\dfrac{4}{5}$ $\tan\theta = -\dfrac{4}{3}$

$\sin 2\theta = 2\sin\theta\cos\theta = 2\left(-\dfrac{4}{5}\right)\left(\dfrac{3}{4}\right) = -\dfrac{24}{20}$

$\cos 2\theta = \cos^2\theta - \sin^2\theta = \left(\dfrac{3}{4}\right)^2 - \left(-\dfrac{4}{5}\right)^2 = -\dfrac{31}{100}$

$\tan 2\theta = \dfrac{2\tan\theta}{1-\tan^2\theta} = \dfrac{2\left(-\dfrac{4}{3}\right)}{1-\left(-\dfrac{3}{4}\right)^2} = -\dfrac{128}{21}$

Using Trig. Power-Reducing Formulas – Ex. 10

Use the Power-Reducing Formulas to prove:
$$\sin^2 x + \cos^2 x = 1$$

$\sin^2 x + \cos^2 x$	$= 1$
$\dfrac{1 - \cos 2x}{2} + \dfrac{1 + \cos 2x}{2}$	
$\dfrac{1 - \cos 2x + 1 + \cos 2x}{2}$	
$\dfrac{1 + 1}{2}$	
$\dfrac{2}{2}$	$= 1$

Using Trig. Power-Reducing Formulas – Ex. 11

Given: $\cos^4 x$, which is a 4th power trig. expression. Rewrite it a 1st power trig. expression.

$\cos^4 x =$	$(\cos^2 x)^2$
	$\left(\dfrac{1 + \cos 2x}{2}\right)^2$
	$\dfrac{1}{4}(1 + \cos 2x)^2$
	$\dfrac{1}{4}[1 + 2\cos 2x + \cos^2(2x)]$
	$\dfrac{1}{4}\left[1 + 2\cos 2x + \left(\dfrac{1 + \cos 4x}{2}\right)\right]$
	$\dfrac{1}{4} + \dfrac{1}{2}\cos 2x + \dfrac{1}{8}(1 + \cos 4x)$
	$\dfrac{1}{2}\cos 2x + \dfrac{1}{8}\cos 4x + \dfrac{3}{8}$

Check with Desmos.
Graph these trig. functions.

$y(x) = (\cos x)^4$

$g(x) = \dfrac{1}{2}\cos(2x) + \dfrac{1}{8}\cos(4x) + \dfrac{3}{8}$

Trig Topics – Part 3

Complex Numbers – Rectangular Form

Complex Numbers – Rectangular Form	
$a + bi$	$a = $ The real part $b = $ The imaginary part

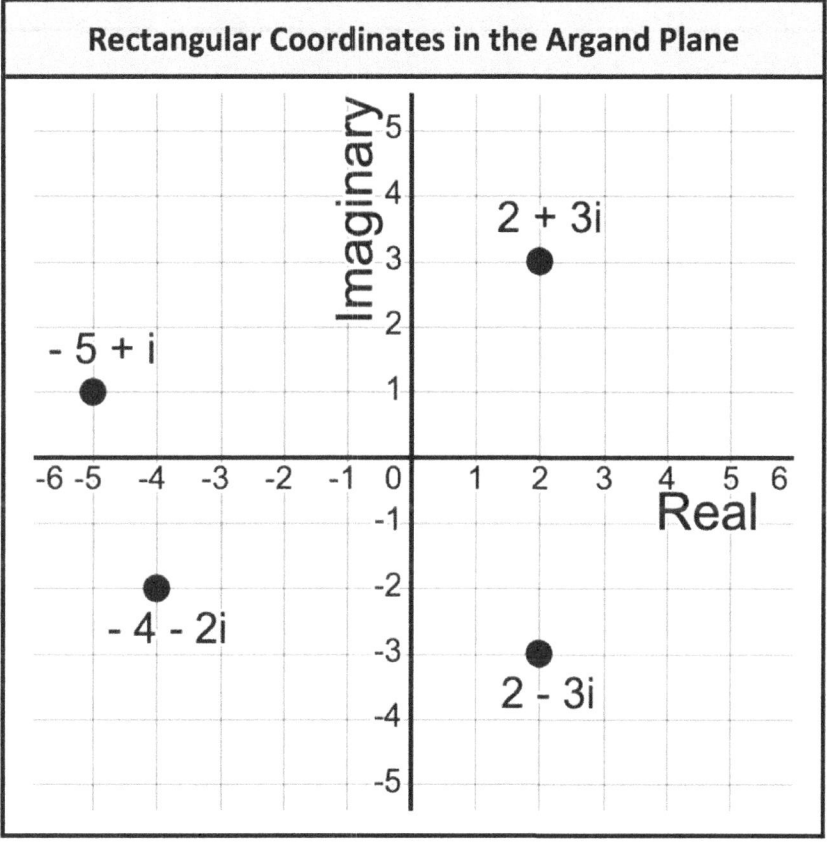

Modulus = Absolute Value
For a complex number: $z = a + bi$ $\|z\|$ = Modulus (or Absolute Value) $\|z\| = \sqrt{a^2 + b^2}$

$z = a + bi$
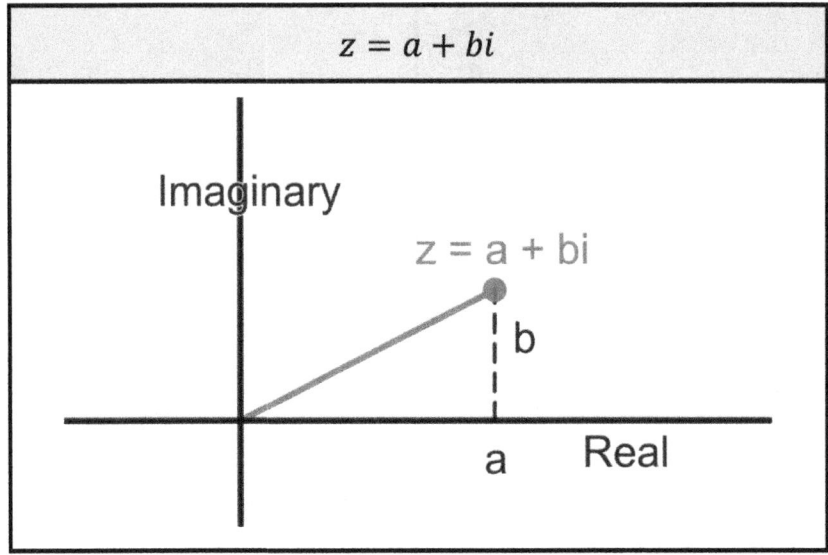

Complex Numbers – Polar Form

Complex Numbers Rectangular Form vs Polar Form	
Rectangular Form	$z = a + bi$
Polar Form	$z = r(\cos\theta + i\sin\theta)$ $z = r\,cis(\theta)$ $z = re^{\theta i}$

Rectangular Coordinates. vs. Polar Coordinates	
Rectangular	(a, b)
Polar	(r, θ)

Definitions
Where: $\quad a = r\cos\theta \qquad\qquad b = r\sin\theta$ $\quad r = |z| = \sqrt{a^2 + b^2} \qquad$ (Modulus) $\quad \theta = \tan^{-1}\left(\dfrac{b}{a}\right) \qquad\qquad$ (Argument)

Polar Coordinates -- Graphing

The Polar Coordinate System is a two-dimensional system used to graph points on a plane. Each point is identified by two polar coordinates (r, θ) where:

r = distance from the origin
θ = the direction

$(r, \theta) = (5, 30°)$	$(r, \theta) = (-5, 30°)$

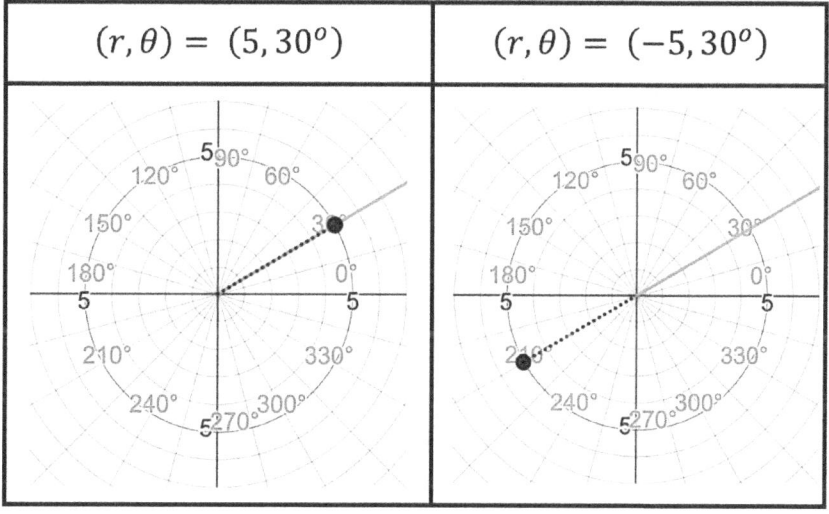

Polar Coordinates -- More Graphing Examples With θ in Degrees

$(r, \theta) = (5, -30°)$	$(r, \theta) = (-5, -30°)$

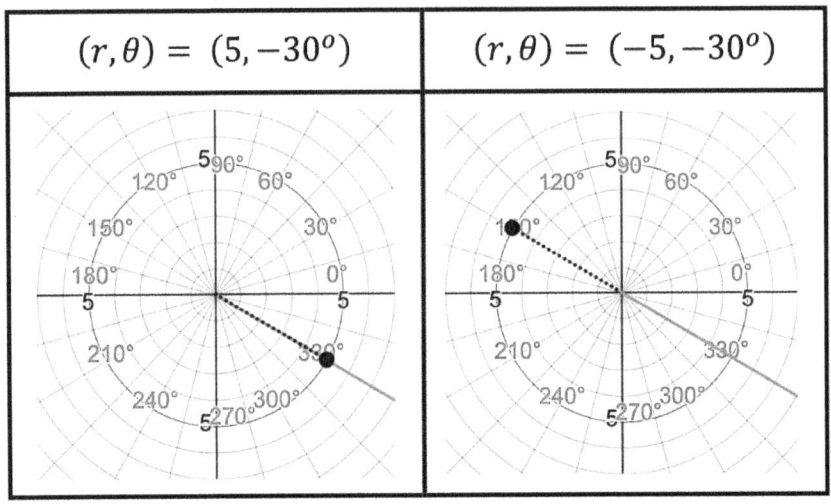

$(r, \theta) = (3, 120°)$	$(r, \theta) = (-3, 120°)$

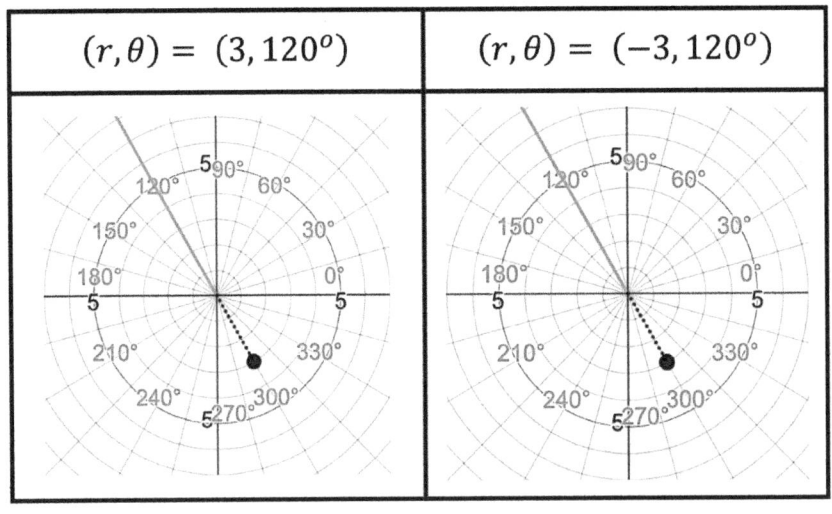

Polar Coordinates -- More Graphing Examples
With θ in Radians   ($\pi \approx 3.1$)

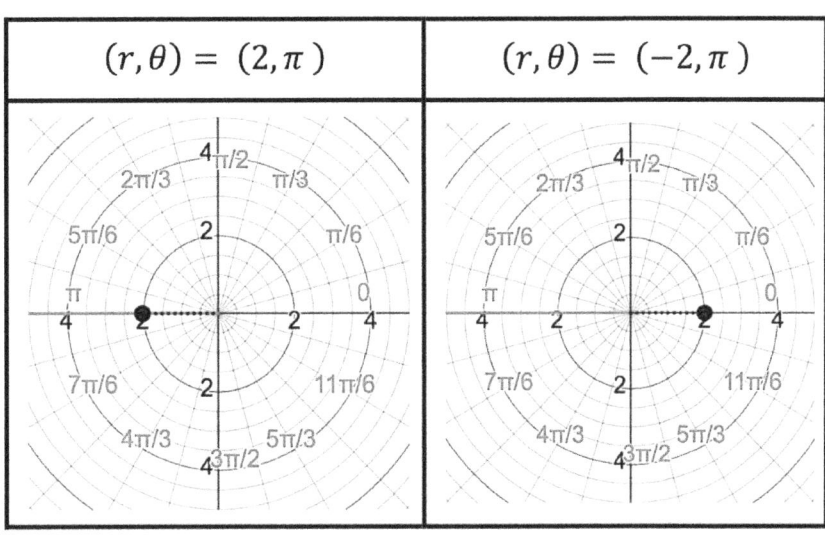

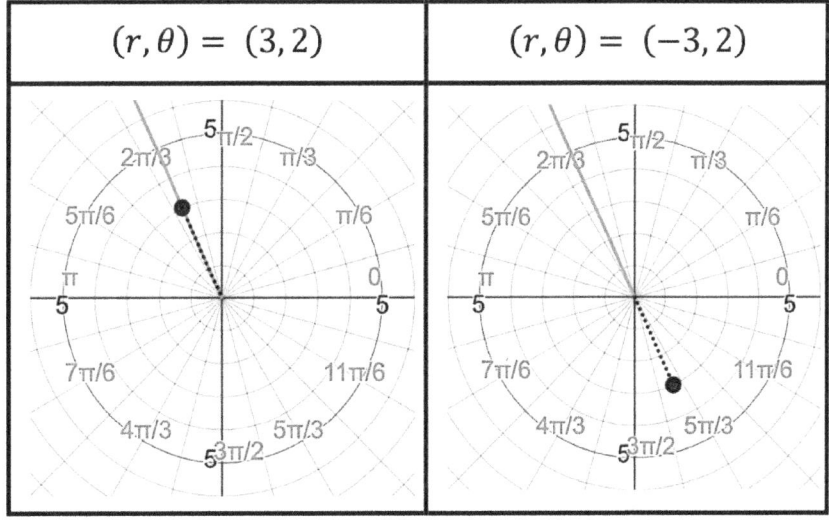

Complex Numbers -- Polar Form – Ex. 1

Write the following complex numbers in Polar Form.

$$z = 1 + i \quad \text{and} \quad z = \sqrt{3} - i$$

$z = 1 + i$

$r = |z| = \sqrt{1^2 + 1^2} = \sqrt{2}$

$\theta = \tan^{-1}\left(\frac{1}{1}\right) = \frac{\pi}{4}$

$z = (r, \theta) = \left(\sqrt{2}, \frac{\pi}{4}\right)$

$z = \sqrt{2}\left(\cos\frac{\pi}{4} + i\sin\frac{\pi}{4}\right)$

$z = \sqrt{2} \operatorname{cis}\left(\frac{\pi}{4}\right)$

$z = \sqrt{3} - i$

$r = |z| = \sqrt{\left(\sqrt{3}\right)^2 + 1^2} = 2$

$\theta = \tan^{-1}\left(\frac{-1}{\sqrt{3}}\right) = -\frac{\pi}{6}$

$z = (r, \theta) = \left(2, -\frac{\pi}{6}\right)$

$z = 2\left(\cos\left(-\frac{\pi}{6}\right) + i\sin\left(-\frac{\pi}{6}\right)\right)$

$z = 2 \operatorname{cis}\left(-\frac{\pi}{6}\right)$

Complex Numbers -- Polar Form – Ex. 2

Graph the following complex numbers on the Polar Coordinate System.

$$z = 1 + i \quad \text{and} \quad z = \sqrt{3} - i$$

$z = 1 + i$	$z = \sqrt{3} - i$
$r = \sqrt{1^2 + 1^2} = \sqrt{2}$ $\theta = \tan^{-1}\left(\frac{1}{1}\right) = \frac{\pi}{4}$	$r = \sqrt{\left(\sqrt{3}\right)^2 + 1^2} = 2$ $\theta = \tan^{-1}\left(\frac{-1}{\sqrt{3}}\right) = -\frac{\pi}{6}$
$(r, \theta) = \left(\sqrt{2}, \frac{\pi}{4}\right)$	$(r, \theta) = \left(2, -\frac{\pi}{6}\right)$

Complex Numbers – Conjugate Pairs

Complex Numbers -- Conjugate Pairs		
Complex Number	$z = a + bi$	$w = c + di$
Conjugate	$\bar{z} = a - bi$	$\bar{w} = c - di$

Properties of Conjugate Pairs			
$\overline{(z + w)}$	$= \bar{z} + \bar{w}$		
$\overline{z\,w}$	$= \bar{z} \cdot \bar{w}$		
$\overline{z^n}$	$= (\bar{z})^n$		
$z \cdot \bar{z}$	$= \|z\|^2$		
$\dfrac{z}{w}$	$= \dfrac{z}{w} \cdot \dfrac{\bar{w}}{\bar{w}}$	$=$	$\dfrac{z \cdot \bar{w}}{\|w\|^2}$

Complex Numbers -- Conjugate Pairs – Ex. 1
Use the Quadratic Equation to find the roots of the equation: $x^2 + 2x + 2 = 0$

$$x = \frac{-b \pm \sqrt{b^2 - 4ac}}{2a}$$

$$x = \frac{-2 \pm \sqrt{2^2 - 4(1)(2)}}{2(1)}$$

$$x = \frac{-2 \pm \sqrt{4 - 8}}{2} = \frac{-2 \pm \sqrt{-4}}{2}$$

$$x = \frac{-2 \pm 2\sqrt{-1}}{2} = \frac{-2 \pm 2i}{2}$$

$$x = -1 \pm i \qquad \text{(Conjugate Pair)}$$

Complex Numbers -- Conjugate Pairs – Ex. 2

Use the Quadratic Equation to find the roots of the equation: $x^2 + x + 1 = 0$

$$x = \frac{-b \pm \sqrt{b^2 - 4ac}}{2a}$$

$$x = \frac{-1 \pm \sqrt{1^2 - 4(1)(1)}}{2(1)}$$

$$x = \frac{-1 \pm \sqrt{1 - 4}}{2} = \frac{-1 \pm \sqrt{-3}}{2}$$

$$x = \frac{-1 \pm \sqrt{3}\,i}{2}$$

$$x = -\frac{1}{2} \pm \frac{\sqrt{3}}{2}i \qquad \text{(Conjugate Pair)}$$

Complex Numbers – Operations

Complex Numbers -- Operations
Operations with complex numbers are usually easier if the number is in Polar Form.

Rectangular Form (a, b)	$z = a + bi$
Polar Form (r, θ)	$z = r(\cos\theta + i\sin\theta)$ $z = r\,cis(\theta)$

Definitions
$z =$ Complex number
$a = r\cos\theta \qquad\qquad b = r\sin\theta$
$r = \|z\| = \sqrt{a^2 + b^2}$ (Modulus)
$\theta = \tan^{-1}\left(\dfrac{b}{a}\right)$ (Argument)

Complex Numbers -- Rectangular Form
Division – Ex. 1

Divide: $(2 + 3i) \div (1 + 4i)$

$(2 + 3i) \div (1 + 4i) = \dfrac{(2 + 3i)}{(1 + 4i)} \cdot \dfrac{(1 - 4i)}{(1 - 4i)}$

$= \dfrac{2 - 8i + 3i - 12i^2}{1^2 - (4i)^2}$

$= \dfrac{2 - 5i - 12(-1)}{1 - 16i^2}$

$= \dfrac{2 - 5i + 12}{1 - 16(-1)}$

$= \dfrac{14 - 5i}{1 + 16} = \dfrac{14 - 5i}{17}$

$= \dfrac{14}{17} - \dfrac{5}{17}i$

Complex Numbers -- Rectangular Form
Multiplication – Ex. 2

Multiply the complex numbers, listed below.

$(a + bi) \cdot (c + di)$
$\quad = ac + adi + bci + bdi^2$

$(2 + 3i) \cdot (1 + 4i)$
$\quad = 2 + 8i + 3i + 12i^2$
$\quad = 2 + 11i + 12(-1)$
$\quad = 2 + 11i - 12 \;=\; -10 + 11i$

$(2 + 3i) \cdot (2 - 3i)$
$\quad = 4 - 6i + 6i - 9i^2$
$\quad = 4 - 9(-1)$
$\quad = 4 + 9 \;=\; 13$

$(0 - i) \cdot (2 - 3i)$
$\quad = -2i + 3i^2$
$\quad = -2i + 3(-1) = -3 - 2i$

Complex Numbers -- Polar Form
Multiplication & Division – Ex. 3

Given:

$$z_1 = (r_1, \theta_1) = r_1(\cos\theta_1 + i\sin\theta_1)$$
$$z_2 = (r_2, \theta_2) = r_2(\cos\theta_2 + i\sin\theta_2)$$

Find: The product and quotient

| $z_1 \cdot z_2$ | $= (r_1 \cdot r_2,\ \theta_1 + \theta_2)$ |

| $\dfrac{z_1}{z_2}$ | $= \left(\dfrac{r_1}{r_2},\ \theta_1 - \theta_2\right)$ |

Complex Numbers -- Polar Form
Multiplication & Division – Ex. 4

Given: $z = (r, \theta) = \left(\sqrt{2}, \dfrac{\pi}{4}\right)$

$w = (r, \theta) = \left(2, -\dfrac{\pi}{6}\right)$

Find: The product $z \cdot w$ and quotient $\dfrac{z}{w}$

$z \cdot w$
$= (r_1 \cdot r_2,\ \theta_1 + \theta_2)$
$= \left(2\sqrt{2},\ \dfrac{\pi}{4} + \left(-\dfrac{\pi}{6}\right)\right)$
$= \left(2\sqrt{2},\ \dfrac{6\pi}{24} - \dfrac{4\pi}{24}\right)$
$= \left(2\sqrt{2},\ \dfrac{\pi}{12}\right)$

$\dfrac{z}{w}$
$= \left(\dfrac{r_1}{r_2},\ \theta_1 - \theta_2\right)$
$= \left(\dfrac{\sqrt{2}}{2},\ \dfrac{\pi}{4} - \left(-\dfrac{\pi}{6}\right)\right)$
$= \left(\dfrac{\sqrt{2}}{2},\ \dfrac{\pi}{4} + \dfrac{\pi}{6}\right)$
$= \left(\dfrac{\sqrt{2}}{2},\ \dfrac{5\pi}{12}\right)$

Complex Numbers – DeMoivre's Thrm.

Complex Numbers -- DeMoivre's Theorem

If $\quad z = (r, \theta)$

$\quad z = r(\cos \theta + i \sin \theta) \; = \; r \, cis \, (\theta)$

Then:

$\quad z^n = (r^n, \, n\theta)$

$\quad z^n = r^n(\cos n\theta + i \sin n\theta)$

$\quad z^n = r^n \, cis(n\theta)$

$\quad n =$ Positive Integer

Complex Numbers -- DeMoivre's Theorem – Ex. 1

Use DeMoivre's Theorem to find: $\left(\frac{1}{2} + \frac{1}{2}i\right)^{10}$

Convert complex number to Polar Form	$r = \sqrt{\left(\frac{1}{2}\right)^2 + \left(\frac{1}{2}\right)^2} = \sqrt{\frac{2}{4}} = \frac{\sqrt{2}}{2}$ $\theta = \tan^{-1}\left(\frac{1/2}{1/2}\right) = \frac{\pi}{4}$ $(r, \theta) = \left(\frac{\sqrt{2}}{2}, \frac{\pi}{4}\right)$ Polar Form
Use DeMoivre's Theorem	$(r, \theta)^n = (r^n, n\theta)$ $\left(\frac{\sqrt{2}}{2}, \frac{\pi}{4}\right)^{10} = \left(\left(\frac{\sqrt{2}}{2}\right)^{10}, 10 \cdot \frac{\pi}{4}\right)$ $\left(\frac{\sqrt{2}}{2}, \frac{\pi}{4}\right)^{10} = \left(\frac{2^5}{2^{10}}, \frac{5\pi}{2}\right) = \left(\frac{1}{32}, \frac{5\pi}{2}\right)$
Expand	$\left(\frac{1}{32}, \frac{5\pi}{2}\right) = \frac{1}{32} \text{cis}\left(\frac{5\pi}{2}\right)$ $= \left(\frac{1}{32}\right)\cos\frac{5\pi}{2} + i\left(\frac{1}{32}\right)\sin\frac{5\pi}{2}$ $= \left(\frac{1}{32}\right)(0) + i\left(\frac{1}{32}\right)(1)$ $= \frac{1}{32}i$

Complex Numbers -- DeMoivre's Theorem – Ex. 2a
Given: $m = 3(\cos 56° + i \sin 56°)$
Find: m^6
Hint: Use DeMoivre's Theorem

Polar Form	$m = 3(\cos 56° + i \sin 56°)$ $m = 3 \, cis \, (56)$	Polar Form
DeMoivre's Theorem	$(r, \theta)^n = (r^n, n\theta)$	
m^6	$= [3 \, cis \, (56)]^6$ $= 3^6 \, cis \, (6 \cdot 56)$ $= 729 \, cis \, (336)$ $= 729 \, (\cos 336 + i \sin 336)$	

Complex Numbers -- DeMoivre's Theorem – Ex. 2b

Given: $m = 3(\cos 56° + i \sin 56°)$

Find: m^6 Use Polar Exponential Form

Polar Form	$m = 3(\cos 56° + i \sin 56°)$ $m = 3\,cis\,(56)$ Polar Form
Polar Exponential Form	$m = 3\,e^{56 \cdot i}$
Polar Exponential Form	$m^6 = \left[3\,e^{56 \cdot i}\right]^6$ $m^6 = 3^6\,e^{6 \cdot 56 i}$ $m^6 = 729\,e^{336 \cdot i}$
Back to Regular Polar Form	$m^6 = 729\,cis\,(336)$ $m^6 = 729\,(\cos 336 + i \sin 336)$

Complex Numbers -- DeMoivre's Theorem – Ex. 3		
Given: $\quad z = \sqrt{2} - \sqrt{2}\, i$		Q4
Find: $\quad z^5$		

r	$= \sqrt{(\sqrt{2})^2 + (\sqrt{2})^2} = \sqrt{4} = 2$
θ	$= \tan^{-1}\left(\frac{-\sqrt{2}}{\sqrt{2}}\right) = \tan^{-1}(-1)$ $= -\frac{\pi}{4} = 2\pi - \frac{\pi}{4} = \frac{7\pi}{4}$ $\qquad$ Q4
Polar Form	$z = 2\,cis\left(\frac{7\pi}{4}\right)$ $\qquad$ Polar Form
z^5	$= \left[2\,cis\left(\frac{7\pi}{4}\right)\right]^5$ $= 2^5\,cis\left(5 \cdot \frac{7\pi}{4}\right) = 32\,cis\left(\frac{35\pi}{4}\right)$ $= 32\,cis\left(\frac{32\pi}{4} + \frac{3\pi}{4}\right) = 32\,cis\left(\frac{3\pi}{4}\right)$ $= 32\left(\cos\frac{3\pi}{4} + i\sin\frac{3\pi}{4}\right)$

Complex Numbers – Roots

Complex Numbers -- Roots

If $\quad z = (r, \theta) = r(\cos\theta + i\sin\theta) = r\,cis(\theta)$

And $\quad n =$ Positive Integer

Then: $\quad z$ has n distinct nth roots.
$$(w_0, w_1, w_2, \ldots w_{n-1})$$

With: $\quad w_k = \left(r^{\frac{1}{n}}, \dfrac{\theta}{n} + \dfrac{2k\pi}{n}\right)$

$\quad\quad\quad k = 0, 1, 2, \ldots n-1$

$\quad\quad\quad\quad\quad\quad\quad n$ distinct roots

Complex Numbers -- Roots – Ex. 1a
Find the six 6th roots of: $z = -8$ And graph the roots on the complex plane.

Convert complex number to Polar Form	$z = -8 + 0i = \sqrt{8^2 + 0^2} = 8$ $\theta = \tan^{-1}\left(\frac{0}{8}\right) = 0$ $z = (r, \theta) = (8, \pi) = 8\,cis(\pi)$
Find the roots.	$w_k = \left(r^{\frac{1}{n}},\ \frac{\theta}{n} + \frac{2k\pi}{n}\right)$ $k = 0, 1, \ldots n - 1$
$r = 8$ $\theta = \pi$ $n = 6$	$w_k = \left(8^{\frac{1}{6}},\ \frac{\pi}{6} + \frac{2k\pi}{6}\right)$ $k = 0, 1, \ldots 5$
$k = 0$	$w_0 = \left(8^{\frac{1}{6}},\ \frac{\pi}{6} + \frac{2(0)\pi}{6}\right) = \left(\sqrt{2},\ \frac{\pi}{6}\right)$
$k = 1$	$w_1 = \left(8^{\frac{1}{6}},\ \frac{\pi}{6} + \frac{2(1)\pi}{6}\right) = \left(\sqrt{2},\ \frac{3\pi}{6}\right)$

Continued …

Complex Numbers -- Roots – Ex. 1b

Find the six 6th roots of: $\quad z = -8$
And graph the roots on the complex plane.

The 6 Roots of $z = -8$

$k = 0$	$w_0 = \left(8^{\frac{1}{6}}, \frac{\pi}{6} + \frac{2(0)\pi}{6}\right) = \left(\sqrt{2}, \frac{\pi}{6}\right)$
$k = 1$	$w_1 = \left(8^{\frac{1}{6}}, \frac{\pi}{6} + \frac{2(1)\pi}{6}\right) = \left(\sqrt{2}, \frac{3\pi}{6}\right)$
$k = 2$	$w_2 = \left(8^{\frac{1}{6}}, \frac{\pi}{6} + \frac{2(2)\pi}{6}\right) = \left(\sqrt{2}, \frac{5\pi}{6}\right)$
$k = 3$	$w_3 = \left(8^{\frac{1}{6}}, \frac{\pi}{6} + \frac{2(3)\pi}{6}\right) = \left(\sqrt{2}, \frac{7\pi}{6}\right)$
$k = 4$	$w_4 = \left(8^{\frac{1}{6}}, \frac{\pi}{6} + \frac{2(4)\pi}{6}\right) = \left(\sqrt{2}, \frac{9\pi}{6}\right)$
$k = 5$	$w_5 = \left(8^{\frac{1}{6}}, \frac{\pi}{6} + \frac{2(5)\pi}{6}\right) = \left(\sqrt{2}, \frac{11\pi}{6}\right)$

Complex Numbers -- Roots – Ex. 1c

Find the six 6th roots of: $z = -8$
And graph the roots on the complex plane.

Graph of the 6 Roots

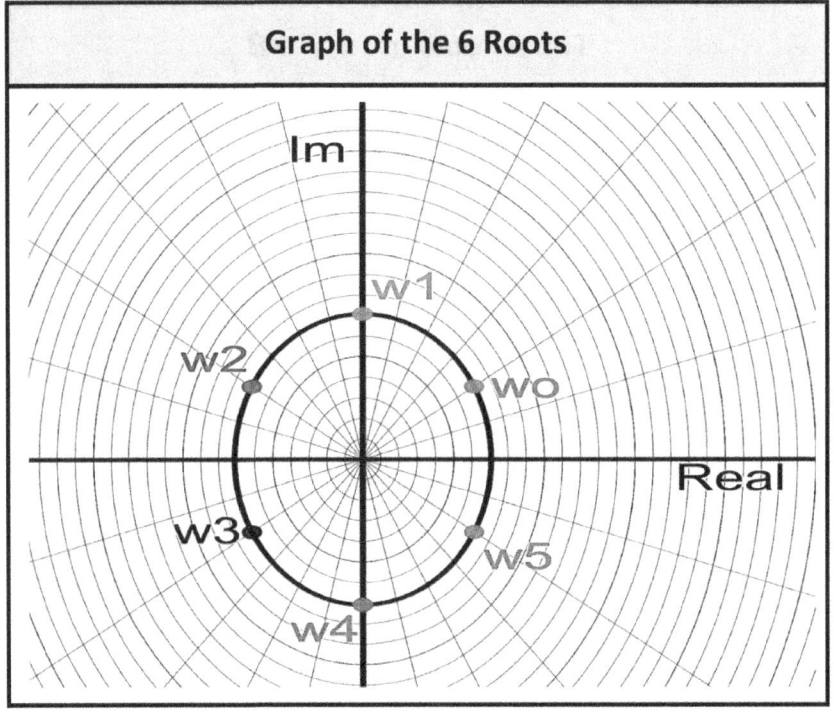

$$|z| = \sqrt{(-8)^2 + 0^2} = 8$$

$$\text{Radius of circle} = 8^{\frac{1}{6}} = \sqrt{2} \approx 1.4$$

	Complex Numbers -- Roots – Ex. 2a
	Given: $z = 16(\cos 100° + i \sin 100°)$ Find the 4th roots of z in polar form.

Complex number is given in Polar Form	$z = 16(\cos 100° + i \sin 100°)$ $z = (r, \theta) = (16, 100°)$ $z = 16\, cis(100°)$
Find the roots.	$w_k = \left(r^{\frac{1}{n}},\ \dfrac{\theta}{n} + \dfrac{2k\pi}{n} \right)$ $\qquad\qquad k = 0, 1, \ldots n-1$
$z^{\frac{1}{4}}$	$= [\,16\, cis(100)\,]^{\frac{1}{4}}$ $= 16^{\frac{1}{4}}\, cis\left(\dfrac{100}{4} + k \cdot \dfrac{360}{4} \right)$ $= 2\, cis(25 + 90 \cdot k)\ ;\quad k = 0, 1, 2, 3$
$k = 0$	$z_0 = 2\, cis(25 + 90 \cdot 0) = 2\, cis(\,25°)$
$k = 1$	$z_1 = 2\, cis(25 + 90 \cdot 1) = 2\, cis(115°)$
$k = 2$	$z_2 = 2\, cis(25 + 90 \cdot 2) = 2\, cis(205°)$
$k = 3$	$z_3 = 2\, cis(25 + 90 \cdot 3) = 2\, cis(295°)$

Complex Numbers -- Roots – Ex. 2b
Given: $z = 16(\cos 100° + i \sin 100°)$ Find the 4th roots of z in polar form.

Previously Found	$z_0 = 2\,cis(25 + 90 \cdot 0) = 2\,cis(25°)$ $z_1 = 2\,cis(25 + 90 \cdot 1) = 2\,cis(115°)$ $z_2 = 2\,cis(25 + 90 \cdot 2) = 2\,cis(205°)$ $z_3 = 2\,cis(25 + 90 \cdot 3) = 2\,cis(295°)$

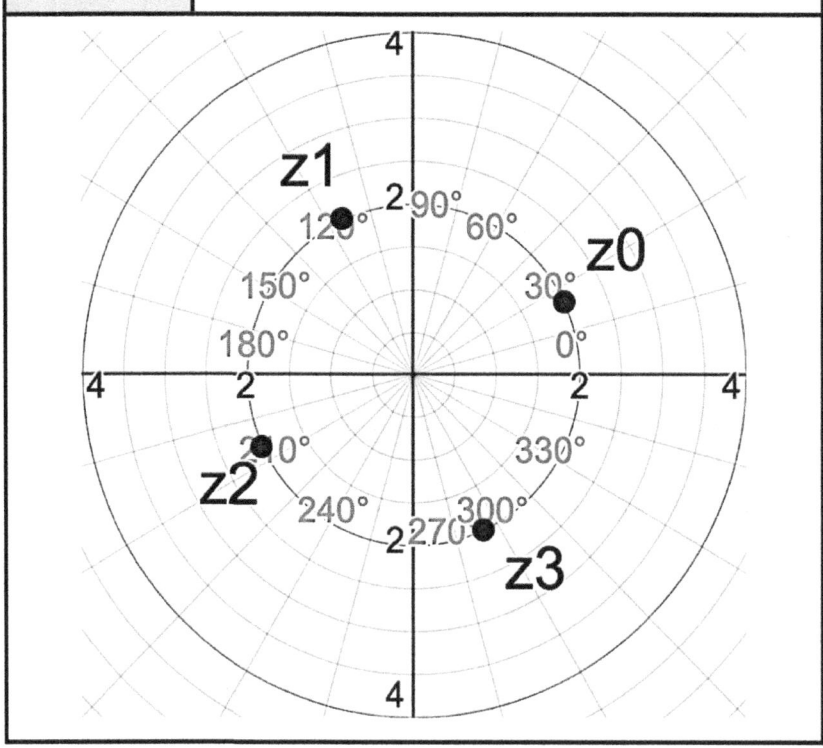

	Complex Numbers -- Roots – Ex. 3a
	Given: $z = -27i$ Find the 3rd roots of z in polar form. Radians.

r	$r = \sqrt{a^2 + b^2} = \sqrt{0^2 + 27^2} = 27$
θ	$\theta = \tan^{-1}\left(\frac{-27}{0}\right) = Undefined$ $\cos\theta = 0 \rightarrow \theta = \frac{\pi}{2}, \frac{3\pi}{2}$ Use $\frac{3\pi}{2}$
Polar Form	$z = r\,cis(\theta) = 27\,cis\left(\frac{3\pi}{2}\right)$
$z^{\frac{1}{3}}$	$= \left[27\,cis\left(\frac{3\pi}{2}\right)\right]^{\frac{1}{3}}$ $= 27^{\frac{1}{3}}\,cis\left(\frac{1}{3} \cdot \frac{3\pi}{2} + k \cdot \frac{2\pi}{3}\right)$ $= 3\,cis\left(\frac{\pi}{2} + \frac{2\pi}{3} \cdot k\right)\,; \quad k = 0, 1, 2$
$k = 0$	$z_0 = 3\,cis\left(\frac{3\pi}{6} + \frac{4\pi}{6} \cdot 0\right) = 3\,cis\left(\frac{3\pi}{6}\right)$
$k = 1$	$z_1 = 3\,cis\left(\frac{3\pi}{6} + \frac{4\pi}{6} \cdot 1\right) = 3\,cis\left(\frac{7\pi}{6}\right)$
$k = 2$	$z_2 = 3\,cis\left(\frac{3\pi}{6} + \frac{4\pi}{6} \cdot 2\right) = 3\,cis\left(\frac{11\pi}{6}\right)$

Complex Numbers -- Roots – Ex. 3b

Given: $z = -27i$

Find the 3rd roots of z in polar form. Radians.

Previously Found

$$z_0 = 3\ cis\left(\frac{3\pi}{6}\right)$$

$$z_1 = 3\ cis\left(\frac{7\pi}{6}\right)$$

$$z_2 = 3\ cis\left(\frac{11\pi}{6}\right)$$

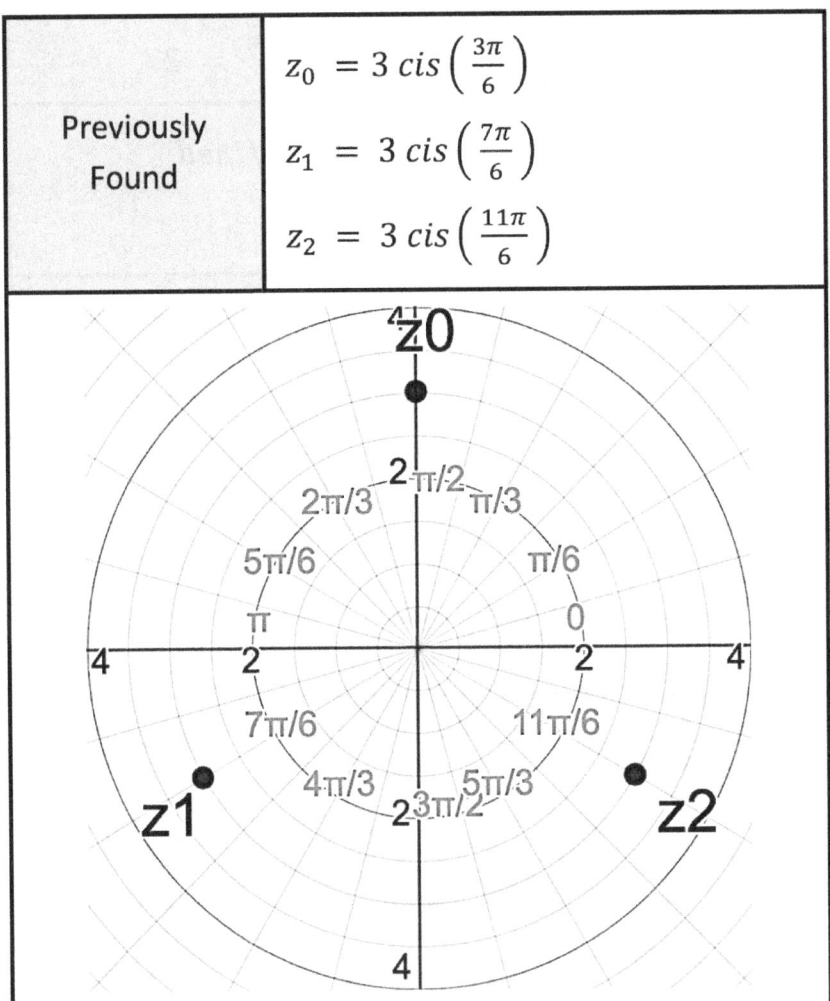

Polar Curves

Polar Curves

A polar equation: $r = f(\theta)$

Note: The radius may change, as θ changes. The value of r depends on the angle, θ

The graph of a polar equation is a **Polar Curve**.
To create a Polar Curve, do the following.

Create a "T" table	θ	$r = f(\theta)$
	0	$r = f(0)$
	30°	$r = f(30)$
	90°	$r = f(90)$
	...	...
Plot points (r, θ) from the table. Then, connect the points. Note the direction.	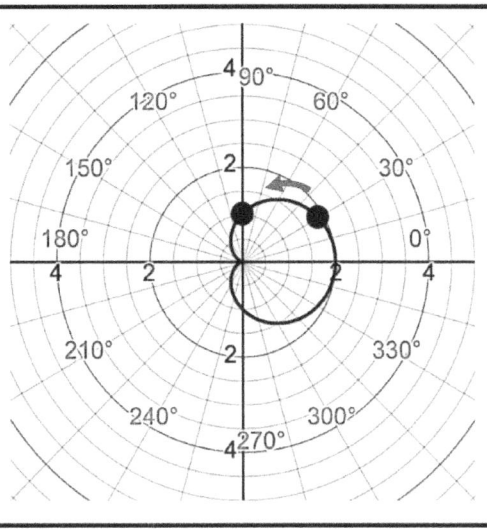	

Polar Curves -- Ex. 1

Sketch the curve represented by $r = 2$

Create a "T" table	θ	$r = 2$
	0	2
	$\frac{\pi}{2}$	2
	π	2
	$\frac{3\pi}{2}$	2
	2π	2

Plot points (r, θ) from the table.

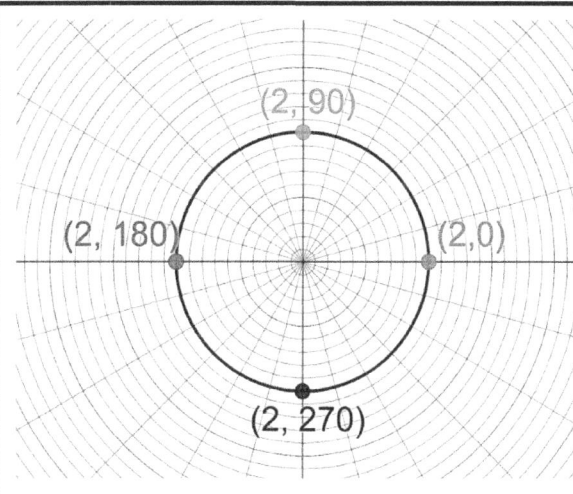

Polar Curves -- Ex. 2

Sketch the curve represented by $\theta = 1$ rad
Note: The angle is constant

Create a "T" table	θ	r = anything
	1	0
	1	1
	1	2
	1	-1
	1	0

Plot points (r, θ) from the table.

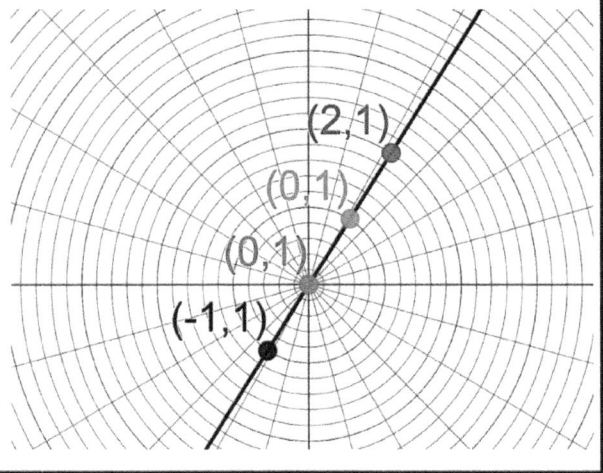

Polar Curves -- Ex. 3

Sketch the curve represented by $r = 2\cos\theta$

Create a "T" table

$r = 2\cos\theta$	θ
2	0
$\sqrt{2}$	$45°$
0	$90°$
-2	$180°$
0	$270°$
2	$360°$

Plot points (r,θ) from the table.

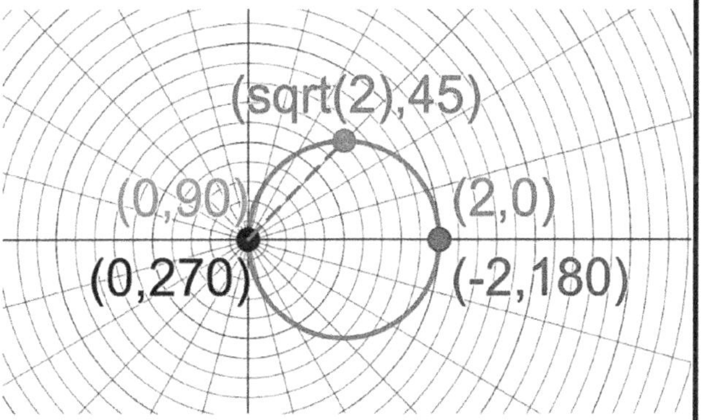

(sqrt(2), 45)
(0, 90)
(2, 0)
(0, 270)
(-2, 180)

Polar Curves -- Ex. 4
Sketch the curve: $r = 1 + \sin\theta$

Create a "T" table	$r = 1 + \sin\theta$	θ
	1	0
	2	$90°$
	1	$180°$
	0	$270°$
	1	$360°$
Plot points (r, θ) from the table.	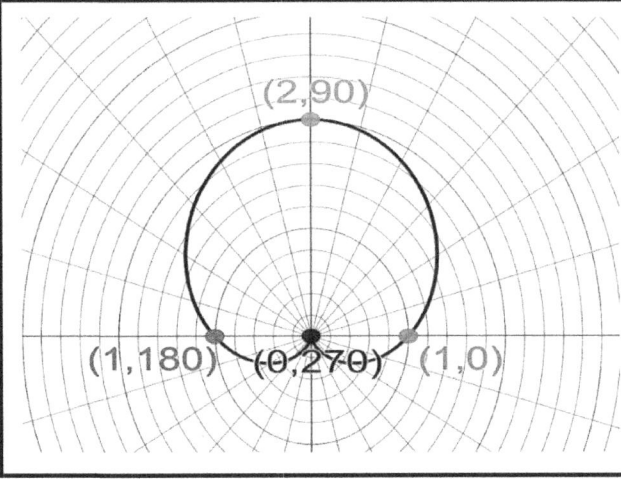	

Polar Curves -- Ex. 5

Sketch the curve: $r = \cos 2\theta$

Create a "T" table

$r = \cos 2\theta$	θ
1	0
0	45°
−1	90°
1	180°
−1	270°

Plot points (r, θ) from the table.

(−1, 270)
(0, 45)
(1, 180)
(1, 0)
(−1, 90)

Polar Curves -- Ex. 6

Sketch the curve: $r = 1 + 3\sin\theta$

Create a "T" table

$r = 1 + 3\sin\theta$	θ
1	0
4	$90°$
1	$180°$
-2	$270°$

Plot points (r, θ) from the table.

Points plotted: (4, 90), (-2, 270), (1, 180), (1, 0)

Polar Curves -- Ex. 7
Sketch the curve: $r = 1 - 3\sin\theta$

Create a "T" table		$r = 1 - 3\sin\theta$	θ
		1	0
		-2	$90°$
		1	$180°$
		4	$270°$
Plot points (r, θ) from the table.	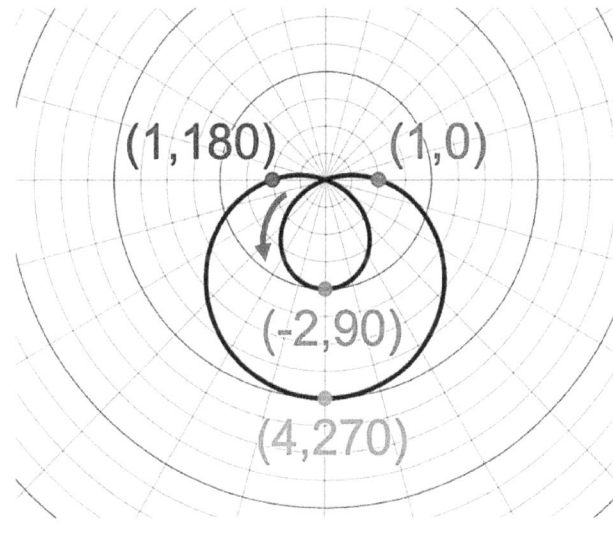		

Polar Curves -- Ex. 8
Sketch the curve: $r = 1 + 3\cos\theta$

Create a "T" table	$r = 1 + 3\cos\theta$	θ
	4	0
	1	$90°$
	-2	$180°$
	1	$270°$
Plot points (r, θ) from the table.	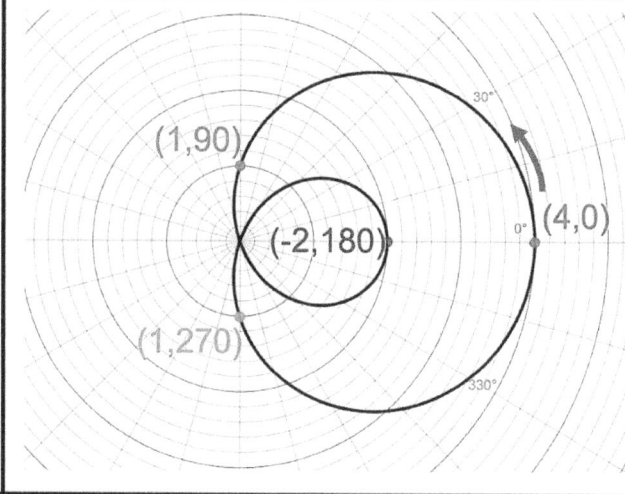	

Polar Curves -- Ex. 9

Sketch the curve: $r = -5$

Create a "T" table	$r = -5$	θ
	-5	0
	-5	$90°$
	-5	$180°$
	-5	$270°$
Plot points (r, θ) from the table.		

	Polar Curves -- Ex. 10
	Sketch the curve: $\theta = \dfrac{\pi}{3}$

Create a "T" table	$r =$ anything		$\theta = \dfrac{\pi}{3}$
	0		$\dfrac{\pi}{3}$
	1		$\dfrac{\pi}{3}$
	-1		$\dfrac{\pi}{3}$
	4		$\dfrac{\pi}{3}$
Plot points (r, θ) from the table.	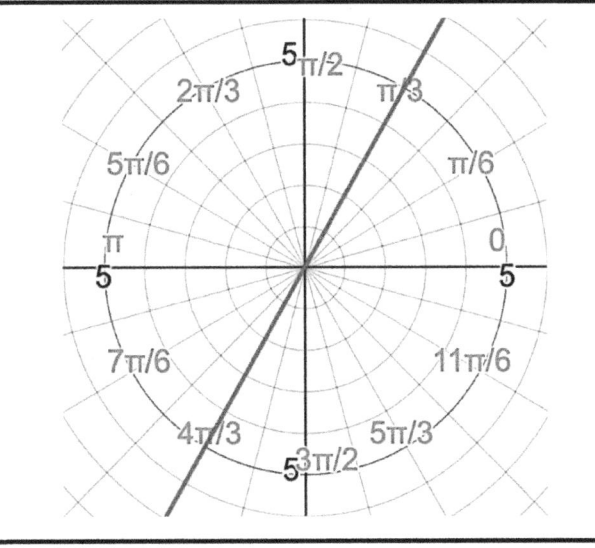

Note: It is a straight line with slope $= \tan\left(\dfrac{\pi}{3}\right) = \sqrt{3}$

Polar Curves -- Ex. 11

Sketch the curve: $r = 5 + 4\cos\theta$

Create a "T" table	$r = 5 + 4\cos\theta$	θ
	9	0
	5	$90°$
	1	$180°$
	5	$270°$
Plot points (r, θ) from the table.		

Note: Lined up along x-axis. Bean shape.

	Polar Curves -- Ex. 12
	Sketch the curve: $r = -5\csc\theta$

Create a "T" table	$r = -5\csc\theta$		θ
	Undefined		0
	-5		$90°$
	Undefined		$180°$
	5		$270°$
Plot points (r,θ) from the table.			
Note:	$r = \dfrac{-5}{\sin\theta}$ $r\sin\theta = -5$ $y = -5$ Horizontal Line		

Polar Curves – Patterns

Polar Curve Patterns
A good way to understand the patterns of polar curves is to use a graphing tool (like Desmos) and try different polar equations.

Desmos has a free APP and a free online graphing tool.

When graphing polar curves with Desmos, click on the tool icon (a wrench) and select the polar grid.

Also select radians or degrees.

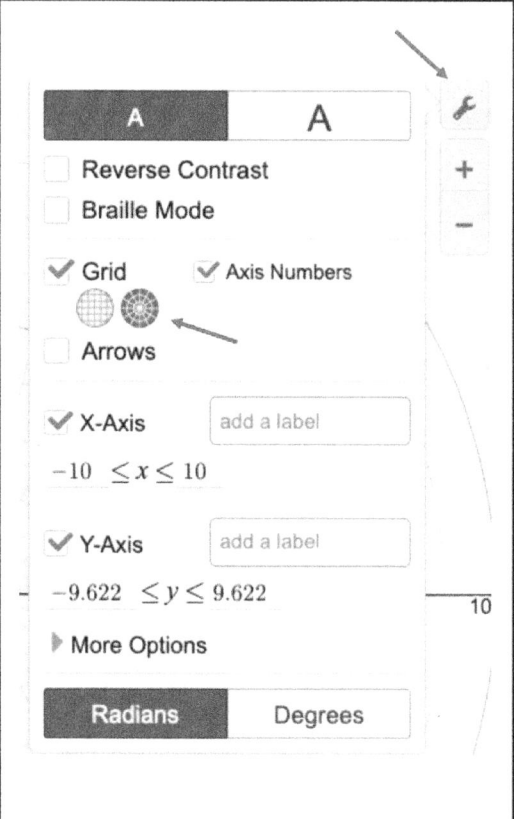

Polar Curve Pattern Overview

Curve	Equation	Example
Limacon	$r = a \pm b \sin \theta$ $r = a \pm b \cos \theta$	
Spiral	$r = a\theta$	
Circles	$r = a \sin \theta$ $r = a \cos \theta$	
Roses	$r = a \sin(n\theta)$ $r = a \cos(n\theta)$	

Polar Curve Pattern Overview – More Examples

Curve	Equation	Example
Circle	$r = 2$	
Spiral	$r = \left(\frac{1}{4}\right)\theta$	
Rose	$r = 2\sin(5\theta)$	
Limacon	$r = 2 + \sin\theta$ Bean	
Limacon	$r = 2 + 2\sin\theta$ Cardioid	
Limacon	$r = 1 + 2\sin\theta$	
Limacon	$r = 2\sin\theta$ Circle	

Polar Curve Pattern – Limacons $r = a + b\sin\theta$		
Here, the curve lines up along the y-axis.		
$b = 0$	Just a constant radius for any θ. A perfect circle with radius = r.	
$a > b$	Perfect circle has a little dent in it. Often called a "bean."	
$a = b$	Perfect circle competes equally with the curving effect of the sine. Called a "heart" or "Cardioid."	
$a < b$	Perfect circle is overwhelmed with curving effect of sine.	

	Polar Curve Pattern – Limacons $r = a + b\cos\theta$	
	Here, the curve lines up along the x-axis.	
$b = 0$	Just a constant radius for any θ. A perfect circle with radius = r.	
$a > b$	Perfect circle has a little dent in it. Often called a "bean."	
$a = b$	Perfect circle competes equally with the curving effect of the cos. Called a "heart" or "Cardioid."	
$a < b$	Perfect circle is overwhelmed with curving effect of cos.	

Polar Curve Pattern – Spiral
$r = n\theta$

$r = \theta$	Starts upward	
$r = -\theta$	Starts downward	
$r = 2\theta$	Starts upward	
$r = -2\theta$	Starts downward	

Polar Curve Pattern – Roses
$r = a\,\text{trig}(n\theta) \qquad n = \text{Even}$
When n is even, $2n =$ the number of leaves

$r = 7\cos(2\theta)$	$r = 7\sin(2\theta)$
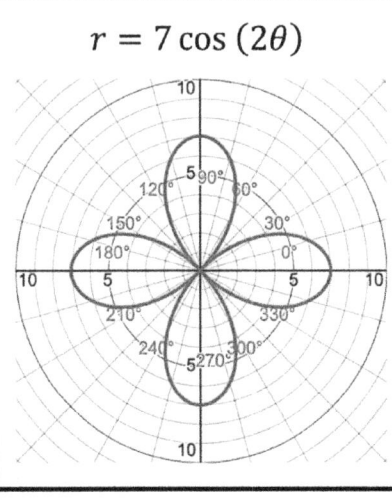	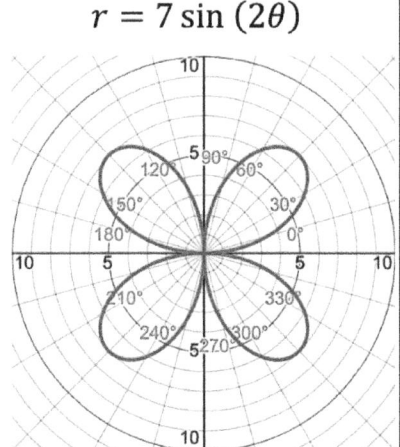
$r = 7\cos(4\theta)$	$r = 7\sin(4\theta)$
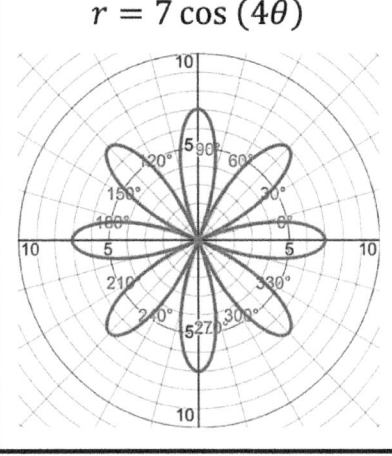	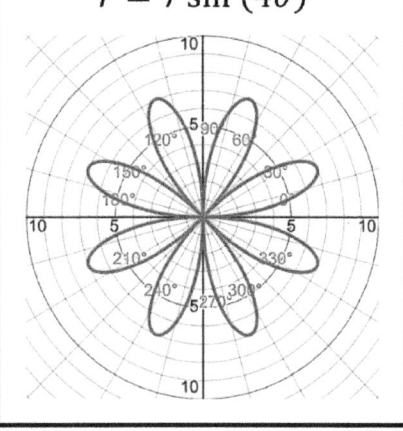

Polar Curve Pattern – Roses
$r = a\,\text{trig}(n\theta) \quad n = \text{Odd}$

When n is odd, $n =$ the number of leaves

$r = 6\cos(3\theta)$	$r = 6\sin(3\theta)$
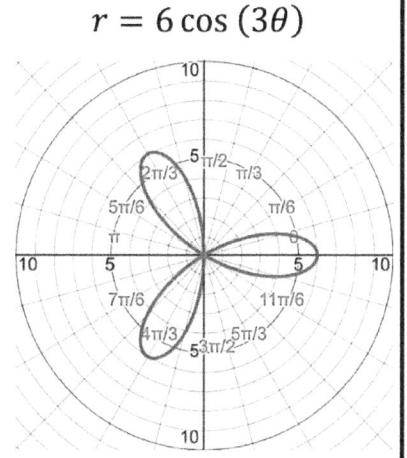	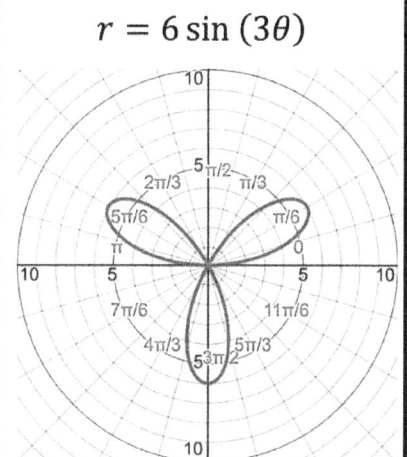
$r = 6\cos(5\theta)$	$r = 6\sin(5\theta)$
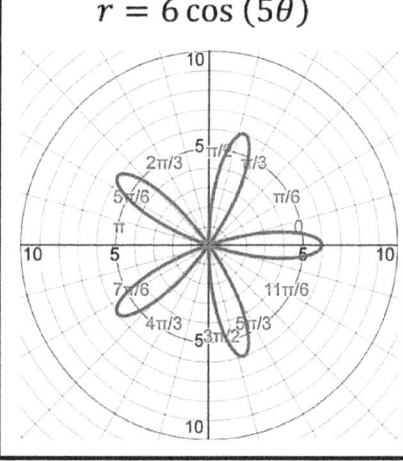	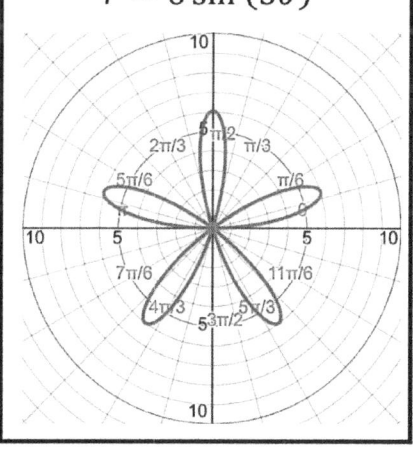

\multicolumn{3}{c}{**Polar Curve Pattern – Roses**}
\multicolumn{3}{c}{$r = a\ \sin(n\theta)$ $n =$ Odd}
n	i^n	$r = \sin(n\theta)$
1	$i^1 = i$ Rose with 1 leaf	
3	$i^3 = (i^2)\,i$ $i^3 = (-1)\,i$	
5	$i^5 = (i^4)\,i$ $i^5 = (1)\,i$	
7	$i^7 = (i^6)\,i$ $i^7 = (-1)\,i$	
9	$i^9 = (i^8)\,i$ $i^9 = (1)\,i$	

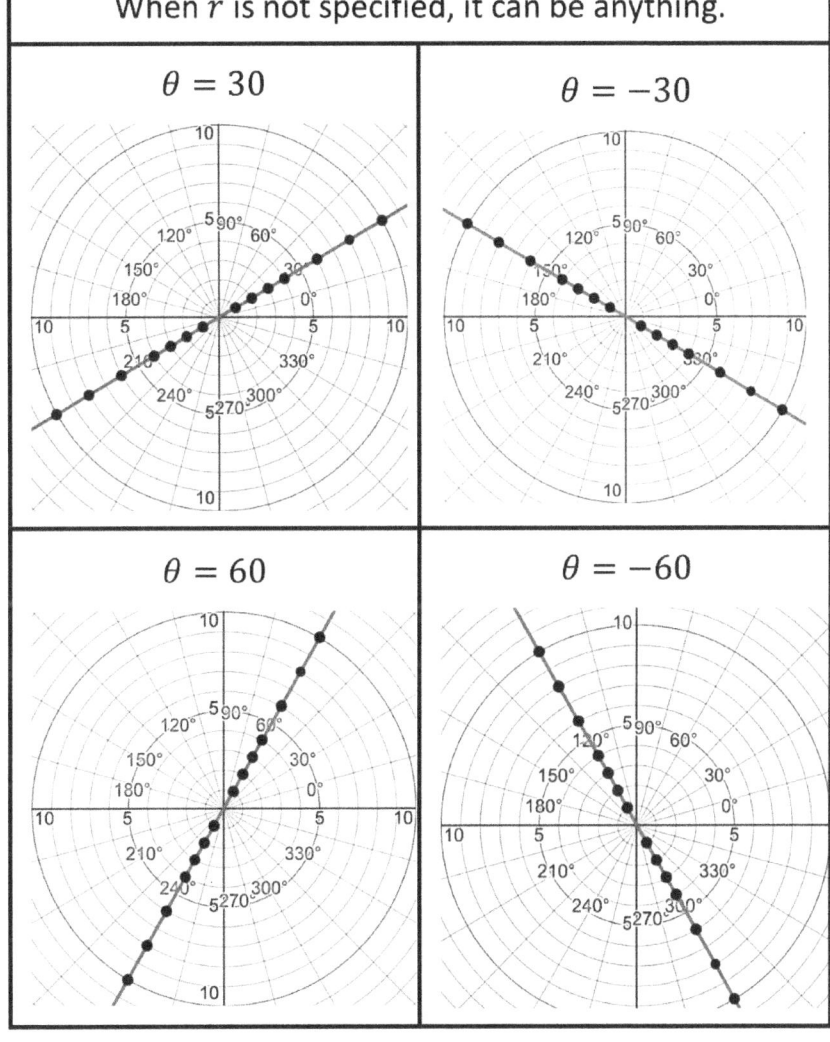

Polar Curves – More Examples

Polar to Rectangular Equation – Ex. 1
Given: $r = -5\sec\theta$ Convert the given polar equation to a rectangular equation.

Look for $r\cos\theta = x$ $r\sin\theta = y$	$r = \dfrac{-5}{\cos\theta}$ $r\cos\theta = -5$ $x = -5$
Notes	This is a vertical line at $x = -5$. Since y is not specified, it can be any value.

Polar to Rectangular Equation – Ex. 2

Given: $r = 3 \cos \theta$

Convert the given polar equation to a rectangular equation.

Look for $r \cos \theta = x$ $r \sin \theta = y$ $r^2 = x^2 + y^2$	$r = 3 \cos \theta$ $r^2 = 3 (r \cos \theta)$ $x^2 + y^2 = 3x$ $x^2 + y^2 - 3x = 0$
Rectangular Equation	$x^2 + y^2 - 3x = 0$

Polar to Rectangular Equation – Ex. 3
Given: $\theta = \frac{\pi}{4}$ Convert the given polar equation to a rectangular equation.

Note:	Here, the angle is constant and $r =$ any value. It's a straight line with slope $= \tan\left(\frac{\pi}{4}\right) = 1$
Look for $r \cos \theta = x$ $r \sin \theta = y$ $r^2 = x^2 + y^2$	$\tan(\theta) = 1$ $\frac{\sin \theta}{\cos \theta} = 1$ $\sin \theta = \cos \theta$ $r \sin \theta = r \cos \theta$ $y = x$ (Rectangular Form)
Rectangular Equation	$y = x$

Polar to Rectangular Equation – Ex. 4

Given: $r = \dfrac{2}{1 - 3\sin\theta}$

Convert the given polar equation to a rectangular equation.

Use: $r\cos\theta = x$ $r\sin\theta = y$ $r^2 = x^2 + y^2$	$r(1 - 3\sin\theta) = 2$ $r - 3r\sin\theta = 2$ $r - 3y = 2$ $r = 2 + 3y$
	$r^2 = (2 + 3y)^2$ $x^2 + y^2 = 4 + 12y + 9y^2$ $x^2 - 8y^2 - 12y - 4 = 0$
Rectangular Equation	$x^2 - 8y^2 - 12y - 4 = 0$

Polar to Rectangular Equation – Ex. 5

Given: $r = \dfrac{2}{3\cos\theta - 4\sin\theta}$; $r = f(\theta)$

Convert the given polar equation to a rectangular equation.

Use: $r\cos\theta = x$ $r\sin\theta = y$ $r^2 = x^2 + y^2$	$r = \dfrac{2}{3\cos\theta - 4\sin\theta}$ $r(3\cos\theta - 4\sin\theta) = 2$ $3r\cos\theta - 4r\sin\theta = 2$ $3x - 4y = 2$ $-4y = 2 - 3x$ $y = -\dfrac{2}{4} + \dfrac{3}{4}x$
Rectangular Equation	$y = -\dfrac{1}{2} + \dfrac{3}{4}x$ $y = f(x)$

Polar to Rectangular Points – Ex. 6

Convert the given polar point to a rectangular point.

Polar (r, θ)	Rectangular (x, y)
$(5, 234°)$	$x = 5 \cos 234 \approx -2.9$ $y = 5 \sin 234 \approx -4.0$ $(x, y) = (-2.9, -4.0)$
$\left(3, -\dfrac{11\pi}{6}\right)$	Use $\theta = -\dfrac{11\pi}{6} + 2\pi = \dfrac{\pi}{6}$ $x = 3 \cos\left(\dfrac{\pi}{6}\right) = \dfrac{3\sqrt{3}}{2}$ $y = 3 \sin\left(\dfrac{\pi}{6}\right) = \dfrac{3}{2}$ $(x, y) = \left(\dfrac{3\sqrt{3}}{2}, \dfrac{3}{2}\right)$

Rectangular to Polar Equation – Ex. 1
Given: $y = 2x - 5$ Convert the given rectangular equation to a polar equation.

Substitute: $x = r \cos \theta$ $y = r \sin \theta$	$y = 2x - 5$ $(r \sin \theta) = 2(r \cos \theta) - 5$
Solve for r	$r \sin \theta - 2r \cos \theta = -5$ $r(\sin \theta - 2 \cos \theta) = -5$ $r = \dfrac{-5}{(\sin \theta - 2 \cos \theta)}$
Polar Equation	$r = \dfrac{-5}{(\sin \theta - 2 \cos \theta)}$ $r = f(\theta)$ That's the goal!

Rectangular to Polar Equation – Ex. 2

Given: $y = \dfrac{\sqrt{3}}{3}x$

Convert the given rectangular equation to a polar equation.

Note:	This is in the form $y = mx$ So, it's a straight line with slope $m = \dfrac{\sqrt{3}}{3}$
Use the slope to identify θ	$\theta = \tan^{-1}\left(\dfrac{\sqrt{3}}{3}\right) = \dfrac{\pi}{6}$
Polar Equation	$\theta = \dfrac{\pi}{6}$
Note:	Since r is not specified, it can be any value.

Rectangular to Polar Equation – Ex. 3
Given: $y^2 = 7x$ Convert the given rectangular equation to a polar equation.

Substitute: $x = r \cos \theta$ $y = r \sin \theta$	$(r \sin \theta)^2 = 7(r \cos \theta)$ $r^2 \sin^2 \theta = 7r \cos \theta$
Solve for r	$r^2 \sin^2 \theta - 7r \cos \theta = 0$ $r(r \sin^2 \theta - 7 \cos \theta) = 0$
	$r = 0$ (Trivial Case) OR $r = \dfrac{7}{\sin^2 \theta}$
Polar Equation	$r = \dfrac{7}{\sin^2 \theta}$ $r = f(\theta)$ That's the goal!

	Rectangular to Polar Point – Ex. 4
	Given: $(x, y) = (-3, -8)$
	Convert the given rectangular point to a polar point.
	Where $r > 0$ and $360° \leq \theta < 720°$.

r	$r = \sqrt{x^2 + y^2}$	
	$r = \sqrt{3^2 + 8^2} = \sqrt{73}$	
θ	$\theta = \tan^{-1}\left(\frac{-8}{-3}\right) = 69.4°$	
	θ must be in original quadrant Q3	
	$\theta = 69.4° + 180° = 249.4°$ in Q3	
	θ must be in specified range.	
	$\theta = 249.4° + 360° = 609.4°$ in Q3	
(r, θ)	$(r, \theta) = (\sqrt{73},\ 609.4°)$	

Rectangular to Polar Equation – Ex. 5

Given: $x + y = 0$

Convert the given rectangular equation to a polar equation.

Substitute: $x = r\cos\theta$ $y = r\sin\theta$	$x + y = 0$ $r\cos\theta + r\sin\theta = 0$ $r(\cos\theta + \sin\theta) = 0$
Solve for r	$r = 0$ (Trivial Case) OR $\sin\theta = -\cos\theta$ $\tan\theta = -1 \quad \rightarrow \quad \theta = \dfrac{3\pi}{4}, \dfrac{7\pi}{4}$
Polar Equation	$\theta = \dfrac{3\pi}{4}$
Note:	Since r is not specified, it can be any value.

Rectangular to Polar Equation – Ex. 6

Given: $y = \dfrac{1}{3}x - 5$

Convert the given rectangular equation to a polar equation.

Simplify	$y = \dfrac{1}{3}x - 5$ $3y = x - 15$
Substitute: $x = r\cos\theta$ $y = r\sin\theta$	$3r\sin\theta = r\cos\theta - 15$
Solve for r	$3r\sin\theta - r\cos\theta = -15$ $r(3\sin\theta - \cos\theta) = -15$ $r = \dfrac{-15}{3\sin\theta - \cos\theta}$
Polar Equation	$r = \dfrac{-15}{3\sin\theta - \cos\theta}$

	Polar Graphs – Ex. 1
	Given: $r = -5$
	Convert the given polar equation to a polar graph.

Note:	The radius is always -5. It is the same for all θ. It is a circle with radius $= 5$
Polar Graph	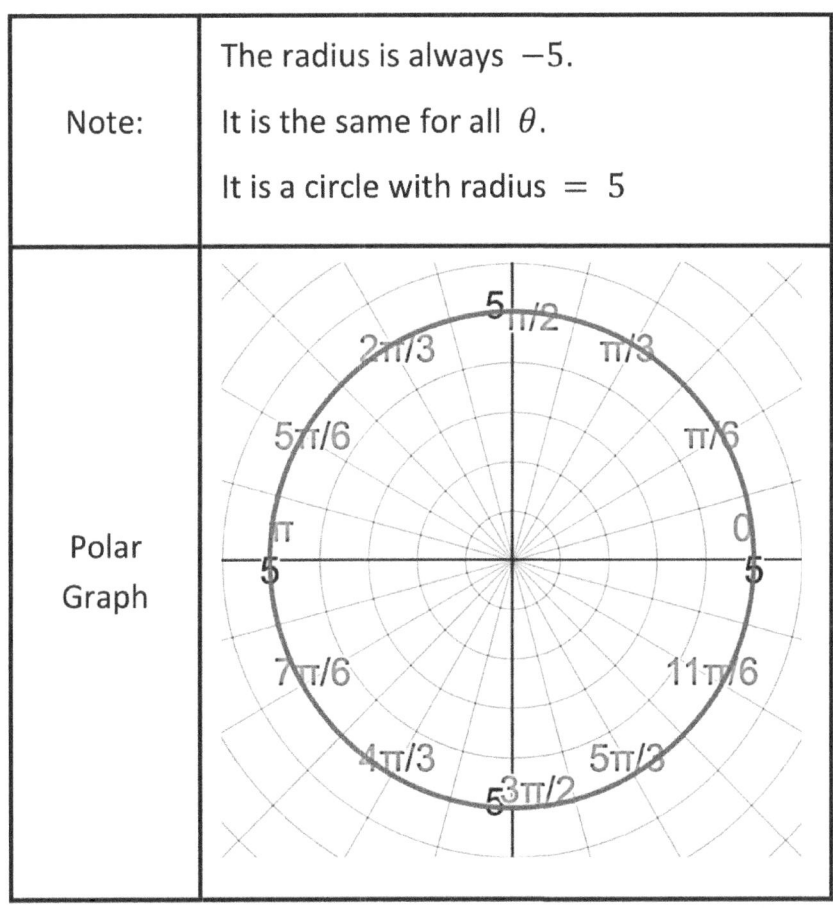

	Polar Graphs – Ex. 2
	Given: $\theta = \dfrac{\pi}{3}$ Convert the given polar equation to a polar graph.

Note:	The angle is always $\theta = \dfrac{\pi}{3}$. Radius is not specified so it can be any value. It is a straight line with Slope $= \tan\left(\dfrac{\pi}{3}\right) = \sqrt{3} \approx 1.7$
Polar Graph	(polar graph showing a straight line through the origin at angle $\pi/3$)

Polar Graphs – Ex. 3

Given: $r = 5 + 4\cos\theta$

Convert the given polar equation to a polar graph.

Note:	Lined up along x axis. Bean shape.			
Create a table of values.	r	9	5	1	5
	θ	0	90	180	270

Polar Graph:

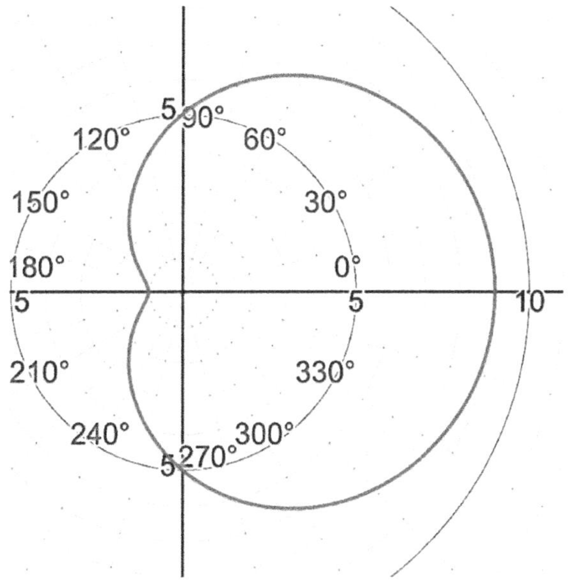

(polar graph showing bean-shaped curve lined up along the x-axis, with points at $r=9, \theta=0°$; $r=5, \theta=90°$; $r=1, \theta=180°$; $r=5, \theta=270°$)

	Polar Graphs – Ex. 4
	Given: $r = -5 \csc \theta$
	Convert the given polar equation to a polar graph.

Rearrange	$r = -5 \csc \theta$ $r = \dfrac{-5}{\sin \theta}$ $r \sin \theta = -5$ $y = -5$ Horizontal Line
Polar Graph	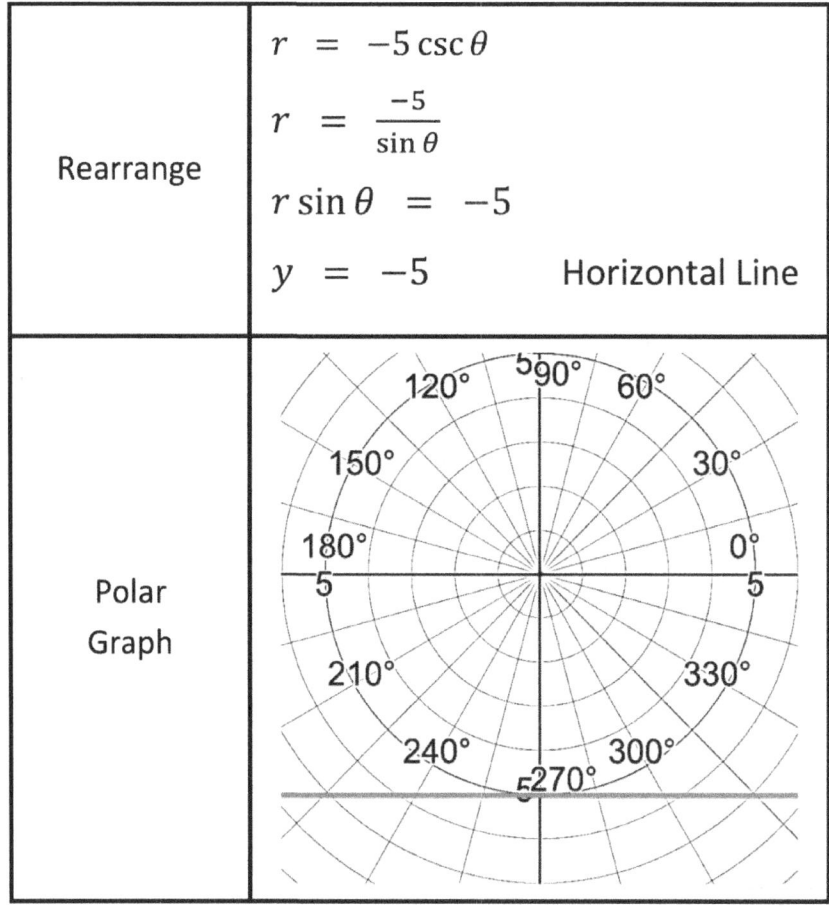

	Cardioid Inner Loop – Ex. 1a
	Given: $r = -\sqrt{3} + 2\sin\theta$
	Find where the inner loop is formed. Radians.

Note	$r = a + b\sin\theta$ With $a < b$ Loop occurs between where $r = 0$
Set $r = 0$	$-\sqrt{3} + 2\sin\theta = 0$ $\sin\theta = \dfrac{\sqrt{3}}{2} \quad \rightarrow \quad \theta = \dfrac{\pi}{3}, \dfrac{2\pi}{3}$
Answer	Inner Loop occurs when: $\dfrac{\pi}{3} \leq \theta \leq \dfrac{2\pi}{3}$
Graph (Extra)	

Cardioid Inner Loop – Ex. 1b

Given: $r = -\sqrt{3} + 2\sin\theta$

Find where the inner loop is formed. Radians.

Answer		
Previously Found	Inner Loop occurs when: $\frac{\pi}{3} \leq \theta \leq \frac{2\pi}{3}$	
Table (Extra)	$r = -\sqrt{3} + 2\sin\theta$	θ
	−1.7	0
	0	$\frac{\pi}{3}$
	1.6	$\frac{\pi}{2}$
	0	$\frac{2\pi}{3}$
	−1.7	π
	−3.5	$\frac{4\pi}{3}$
	−3.7	$\frac{3\pi}{2}$

Polar Curves Intersection – Ex. 1a

Given: $r_1 = 4\sin 2\theta$ and $r_2 = 4\cos\theta$

Find Simultaneous Solution(s)

Make a Sketch	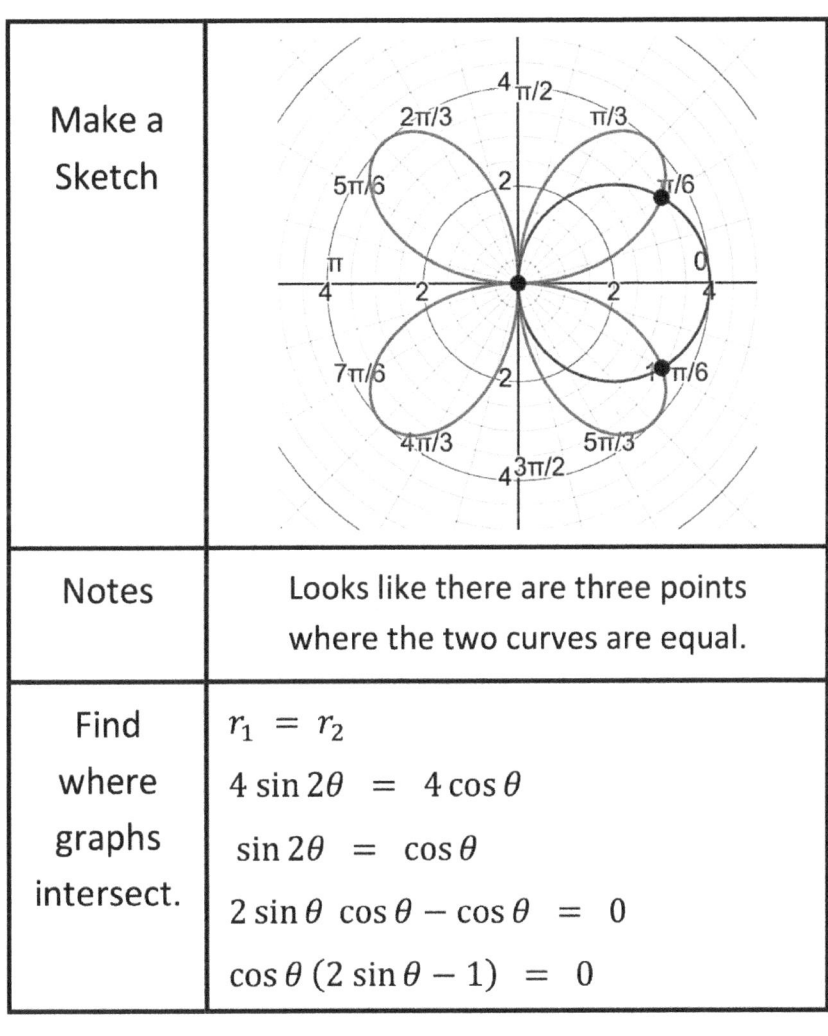
Notes	Looks like there are three points where the two curves are equal.
Find where graphs intersect.	$r_1 = r_2$ $4\sin 2\theta = 4\cos\theta$ $\sin 2\theta = \cos\theta$ $2\sin\theta\cos\theta - \cos\theta = 0$ $\cos\theta(2\sin\theta - 1) = 0$

Polar Curves Intersection – Ex. 1b

Given: $r_1 = 4\sin 2\theta$ and $r_2 = 4\cos\theta$

Find Simultaneous Solution(s)

Previously Found	Intersection when: $r_1 = r_2$ $\cos\theta\,(2\sin\theta - 1) = 0$	
Solve for θ	$\cos\theta = 0$ $\theta = \dfrac{\pi}{2}, \dfrac{3\pi}{2}$	$2\sin\theta = 1$ $\sin\theta = \dfrac{1}{2}$ $\theta = \dfrac{\pi}{6}, \dfrac{5\pi}{6}$

θ	(r,θ)	$x = r\cos\theta$	$y = r\sin\theta$
$\dfrac{\pi}{2}$	$\left(0, \dfrac{\pi}{2}\right)$	0	0
$\dfrac{3\pi}{2}$	$\left(0, \dfrac{3\pi}{2}\right)$	0	0
$\dfrac{\pi}{6}$	$\left(2\sqrt{3}, \dfrac{\pi}{6}\right)$	3	$\sqrt{3} \approx 1.7$
$\dfrac{5\pi}{6}$	$\left(-2\sqrt{3}, \dfrac{5\pi}{6}\right)$	3	$-\sqrt{3}$

The 3 intersection points are:

$(x,y) = (0,0),\ (3,\sqrt{3}),\ (3,-\sqrt{3})$

References

| **References** |

- Algebra and Trigonometry, Structure and Method, Book 2, Houghton Mifflin, Richard Brown, Mary Dolciani, Robert Sorgenfrey, Robert Kane, 1992.

- Mathematics, Structure and Method, Course 2, Mary Dolciani, Robert Sorgenfrey, John Grahm, McDougal Littell, 2001.

- Precalculus, A Graphing Approach, 2nd Edition, Larson, Hostetler, Edwards, Houghton Mifflin Company, 1997.

- Essentials of College Algebra, Richard Aufmann, Richard Nation, Houghton Mifflin, 2006.

- One-Page Summaries for Algebra, Geometry, and Pre-Calculus, Kathryn Paulk, 2023.

- Complex Numbers and Polar Curves for Pre-Calc and Trig: With Problems and Detailed Solutions, Kathryn Paulk, 2023.

Other Books by Kathryn Paulk

Other Books by Kathryn Paulk

- Algebra 1 Help
- Algebra 2 Help
- Pre-Calculus and Trig Help
- College Algebra Help

- Calculus 1 Review in Bite-Size Pieces
- Calculus 2 Review in Bite-Size Pieces
- Calculus 3 Review in Bite-Size Pieces
- Differential Equations With Applications: Class Notes With Examples

- One-Page Summaries for Algebra, Geometry, and Pre-Calculus
- Graphing Functions Using Transformations for Algebra & Pre-Calc.
- Complex Numbers and Polar Curves For Pre-Calc and Trig: With Problems and Detailed Solutions
- Discrete and Continuous Probability Distributions: A Creative Comparison (V2)

- Teach Your Child to SWIM

BIG MATH For Little Kids

Workbooks for Young Children & Solution Manuals for Parents

- Introduction to Numbers

- Introduction to Fractions
  by Sharing Things

- Introduction to Counting and Fractions
  by Cooking Breakfast

- Learn About Fractions   *****
  by Baking Cookies

- Adding Big Numbers, Guessing Numbers
  and Secret Codes

- Learn to Graph by Riding Bikes
  on Graph Paper

550

Made in the USA
Middletown, DE
27 February 2025

71972216R00305